John Stuart Jackson

Geometrical Conic Sections

An Elementary Treatise in which the Conic Sections Are Defined as the Plane Sections of a Cone and Treated by the Method of Projections

John Stuart Jackson

Geometrical Conic Sections
An Elementary Treatise in which the Conic Sections Are Defined as the Plane Sections of a Cone and Treated by the Method of Projections

ISBN/EAN: 9783337251758

Printed in Europe, USA, Canada, Australia, Japan

Cover: Foto ©berggeist007 / pixelio.de

More available books at **www.hansebooks.com**

GEOMETRICAL CONIC SECTIONS.

AN ELEMENTARY TREATISE

IN WHICH THE CONIC SECTIONS ARE DEFINED AS THE PLANE SECTIONS
OF A CONE, AND TREATED BY THE METHOD OF PROJECTIONS.

BY

J. STUART JACKSON, M.A.,
LATE FELLOW OF GONVILLE AND CAIUS COLLEGE, CAMBRIDGE.

London and New York:
MACMILLAN AND CO.
1872.

[All Rights reserved.]

Cambridge:
PRINTED BY C. J. CLAY, M.A.
AT THE UNIVERSITY PRESS.

PREFACE.

THE following pages have been written with a view to give the student the benefit of the Method of Projections as applied to the Ellipse and Hyperbola. This Method is calculated to produce a material simplification in the treatment of those curves and to make the proof of their properties more easily understood in the first instance and more easily remembered. It is also a powerful instrument in the solution of a large class of problems relating to these curves.

When the Method of Projections is admitted into the treatment of the Conic Sections there are many reasons why they should be defined, not as has been the case of late years with reference to the focus and directrix, but according to the original definition from which they have their name, as plane Sections of a Cone. First and principally, because this definition gives an immediate proof of the relation by projection of the Ellipse and Circle and of the General and Rectangular Hyperbolas: and in the second place, it naturally divides the properties that may be proved by projection from those connected with the focus and directrix, and thus introduces a valuable simplification into the treatment of the subject. It is also a consideration of some importance that we can see at once from the form of the cone, the general form of the curves that may be cut from it by a plane in different positions; and, by turning the plane about a certain line, we see how the curves pass from one form into another.

It is hoped these may be thought sufficient reasons for departing from the definition which has been in use of late years.

I have retained the proof of the constancy of the ratio of the rectangles under the segments of intersecting chords in constant directions (Chap. v. § 15), given by Dean Hamilton in his treatise published in 1773, as an improving and interesting study and one that is in harmony with the scheme of this work.

I have to acknowledge the great kindness of Rev. Chas. Taylor, Fellow of St John's College, Cambridge, in permitting me the use of his excellent proof of the intersection of tangents at the extremities of a chord of any Conic Section on the diameter of the chord; with the proofs of $CV.CT = CP^2$ (Chap. v. § 8), and $PV = PT$ (Chap. vii. § 20), depending on it: and also of his proof of $SQP = HQP'$ (Chap. iv. § 13).

Professor Adams' property of the tangent is indispensable in any geometrical treatment of the Conic Sections, and I have his kind permission to make use of it.

I shall be greatly obliged by any corrections and hints towards improvement in case this work should be so fortunate as to reach a second edition.

9, BROOKSIDE, CAMBRIDGE,
Christmas, 1871.

TABLE OF CONTENTS.

CHAPTER I.

ON THE METHOD OF PROJECTIONS.

ART.   PAGE
1. Definitions 1
2. Construction 1
3—5. Relation between lines and their projection . . . 2
6. Effect of projection on angles 4
7.   ,,   ,,   ,,   areas 5
8. Tangents project into tangents 5
9. Object of the Method 6

CHAPTER II.

ON THE CONE AND SPHERE.

1, 2. Definitions, &c. 7
3. Every plane section of a sphere is a circle 8
4, 5. Tangents to a sphere from an external point are equal . 8

CHAPTER III.

ON THE ELLIPSE.

1. Definition 10
2. The ellipse is the projection of a circle 11
3. Minor axis 12
4. Conjugate diameters 13
5. Supplemental chords 14
6. $QV^2 : CD^2 :: PV . Vp : CP^2$ 14

ART.

7—9. $CV.CT = CP^2$; $CN.CT = CA^2$, &c. 14
10. $PF.PG = BC^2$; $PF.Pg = AC^2$ 15
11. $NG : NC :: BC^2 : AC^2$ 16
12. $CP^2 + CD^2 = AC^2 + BC^2$ 16
13. $QO.Oq : RO.Or$ a constant ratio 17
14. Circumscribing parallelogram 18

CHAPTER IV.

FOCAL PROPERTIES OF THE ELLIPSE.

1,2. Preliminary 24
3—5. Foci and directrices 25
6. Secant and tangent 29
7. Tangent equally inclined to the focal distances . . . 30
8. Auxiliary circle 31
9. $SY.HZ = BC^2$ 32
10. $SP.HP = CD^2$ 33
11. $ST : QN :: AS : AX$ 33
12—14. Intersecting tangents 34
15. $SG : SP :: SA : AX$ 36

CHAPTER V.

ON THE HYPERBOLA.

1. Definition 42
2. Projection of the rectangular hyperbola 43
3. Magnitude of the axes 46
Properties of the rectangular hyperbola:—
    4. Asymptotes 47
    5. Intercepts between asymptotes and curve equal. Locus of the
        middle points of parallel chords 48
    6. $QV^2 + CP^2 = VU^2$ 51
    7,8. $CV.CT = CP^2$; $CN.CT = CA^2$ 52
    9. Conjugate hyperbola, conjugate diameters . . . 54
    10. Inscribed parallelogram 56
    11. Intercepts on a line parallel to an asymptote . . 57

ART.		PAGE
12, 13.	These properties transferred to the general hyperbola	58
14.	Supplemental Chords	61
15, 16.	Ratio of rectangles on the segments of intersecting chord	61
17.	$NG : NC :: BC^2 : AC^2$	65
18.	$PF \cdot PG = BC^2$; $PF \cdot Pg = AC^2$	65
19.	Asymptotes parallel to two generating lines of the cone	66

CHAPTER VI.

FOCAL PROPERTIES OF THE HYPERBOLA.

(See contents of Chapter IV.)

CHAPTER VII.

ON THE PARABOLA.

1.	Definition	87
2.	Limiting form of semi-ellipse or semi-hyperbola	88
3.	$NP^2 = 4AS \cdot AN$	89
4.	$QR \cdot Q'R = 4AS \cdot PR$	90
5.	Locus of middle points of parallel chords	91
6.	$QR^2 = 4AS \cdot PV$	92
7—9.	Focus and directrix	93
10.	Secant and tangent	96
11.	Tangent equally inclined to the focal distance and the axis of the curve	96
12.	Tangents at the extremities of a focal chord	97
13.	Perpendicular from the focus on the tangent	97
14—18.	Intersecting tangents	98
19.	Tangents at the extremities of a chord intersect on its diameter	99
20.	$PT = PV$	99
21.	$QV^2 = 4SP \cdot PV$	100
22.	Length of focal chord	101
23.	Ratio of the rectangles under segments of intersecting chords	101
24, 25.	The normal	102

ERRATA.

Two lines from the bottom of p. 61, after *rectangles* insert *under the segments*.

p. 72, *for* Chap. VII. *read* Chap. VI.

p. 87, *for* Chap. VIII. *read* Chap. VII.

CONIC SECTIONS.

CHAPTER I.

On the Method of Projections.

1. DEFINITIONS. The *Orthogonal projection of a point* on a plane is the foot of the perpendicular drawn from the point to the plane.

The plane on which the projection is made is called the *plane of projection*.

The *orthogonal projection of a line* on a plane is the line traced on the plane by a straight line which passes through all successive points of the given line and is always perpendicular to the plane of projection.

In the present treatise when projections are spoken of it may be understood that orthogonal projections are intended.

If a straight line be drawn through two adjacent points of a curve, and the two points move towards one another on the curve until they coincide, the straight line in its final position is said to be a *tangent* to the curve at the point which it has finally in common with it.

2. In order to project a straight line AB (fig. § 5) on a plane abc, we must draw the perpendicular Aa from some

point A of AB to the plane of projection (Euclid XI. 11), then make a plane pass through AB, Aa; it will be perpendicular to the plane of projection (XI. 18). Hence the perpendicular drawn from any point P of AB to ab the intersection of the planes will be perpendicular to the plane of projection; and the straight line ab will be the projection of AB. Hence the projection of a straight line is also a straight line.

3. The projections of parallel straight lines are also parallel.

From points A and C (fig. § 5) in the parallel straight lines AB, CD draw Aa, Cc perpendicular to the plane of projection, these lines will be parallel (Euclid XI. 6), and the planes BAa, DCc will be parallel (XI. 15): also the plane of projection will cut these planes in parallel lines ab, cd (XI. 16), and these are the projections of AB, CD. Hence the projections of parallel straight lines are parallel.

4. A line of finite length and its projection are cut in the same ratio by any point and its projection.

Let a, p, b be the projections of the points A, P, B in the straight line AB (fig. § 5), draw AE perpendicular to Bb cutting Pp in Q, then Ap, Qb are parallelograms, and AQ, QE equal to ap, pb. Also PQ is parallel to BE the side of the triangle BAE, and therefore $AQ : QE :: AP : PB$, or $ap : pb :: AP : PB$. The line AB and its projection ab are cut in the same ratio by the point P and its projection p.

5. Parallel straight lines of finite lengths are diminished by projection in the same ratio.

Let the parallel straight lines AB, CD of finite length have projections ab, cd. From A and C draw AE, CF perpendicular to Bb, Dd: these lines are equal to ab, cd: they are also parallel to these parallel straight lines and are therefore parallel to one another (Euclid XI. 9). The lines AB, CD are also parallel, hence the angles BAE, DCF are equal

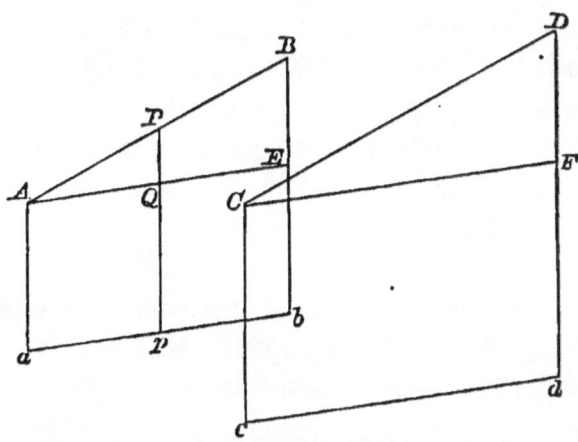

(XI. 10): the angles at E and F are right angles, and the triangles BAE, DCF are similar. Hence

$$BA : AE :: DC : CF \text{ or } BA : ab :: DC : cd;$$

AB, CD are diminished by projection in the same ratio.

Hence it appears that the ratio in which any line is altered by projection from one plane to another depends only upon its inclination to the line of intersection of the planes. It may be seen that lines parallel to the line of intersection are unaltered by projection, and that lines more inclined to it are diminished in a greater ratio than those less so: while lines at right angles to the line of intersection are diminished in the greatest ratio. The ratio which a line at right angles to the line of intersection of its plane with the plane of projection bears to its projection may conveniently be called the *maximum ratio* for the two planes, and it will increase from unity to a ratio as great as we please when we increase the inclination of the planes from zero to a right angle.

In other words, a line at right angles to the line of intersection may have its projection as nearly equal to itself or as small as we please by taking the plane of projection at first very slightly inclined to the plane of the line, and then increasing the inclination till it becomes a right angle. The projection is the base of a right-angled triangle of which the line is the hypotenuse, and the inclination of the planes the angle at the base. By increasing the angle at the base from zero to a right angle we diminish the base from the length of the hypotenuse to zero.

6. Angles will be in general increased or diminished in magnitude by projection, except those contained by lines parallel and right angles perpendicular to the line of intersection.

Of the four angles contained by two intersecting lines, that which is subtended by the line of intersection and the vertical opposite angle will be increased, the other two diminished by projection; and these angles may be increased and diminished to any extent by making the inclination of the plane of projection sufficiently great. If, for example, it be required to increase the acute angle ACB subtended by AB the line of intersection to a right angle; on AB describe the semicircle AcB, and draw CD perpendicular to AB cutting the semicircle in c.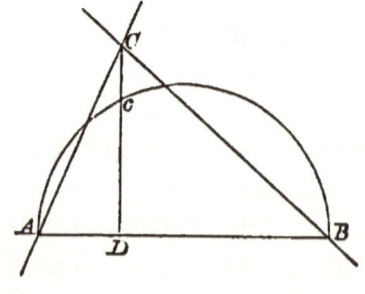
Then let the inclination be so chosen that the maximum ratio shall be $CD : cD$: then CD will have its projection equal to cD, and the angle ACB will project into an angle equal to AcB, i.e. to a right angle.

This only holds, however, when D lies between A and B. If D fell outside AB, the angle ACB might be increased or diminished by projection, and would be greatest when the circle circumscribing ABC touched CD, and in that case DC would be a mean proportional between DA and DB.

7. The area of any figure will be reduced by projection in the maximum ratio of the planes.

For every triangle may be divided by a line through one of the angular points parallel to the line of intersection into two, each having its base parallel and its altitude at right angles to the line of intersection: the base of each will be unaltered by projection, the altitude diminished in the maximum ratio. Hence the area of each and therefore of the two together will be diminished in the maximum ratio. Hence, whatever its magnitude or position, every triangle is diminished by projection in the maximum ratio of the planes.

But every figure, whether rectilinear or curvilinear, may be as nearly occupied by triangles as we please, by making them sufficiently numerous. Each of these will be diminished by projection in the maximum ratio of the planes; and therefore the figure composed of them will be diminished in the same ratio.

8. The projection of the tangent to a curve at any point is the tangent to the projection of the curve at the projection of the point.

If p, q be the projections of two points P and Q of a curve, the straight line pq is the projection of the straight line PQ; and if p, q move so as to be always the projections of P and Q as they approach one another on the curve until they coincide and Q is merged in P, p and q will also approach one another on the projected curve, and will finally coincide in the projection of P, PQ then becomes the tangent to the curve at P, and its projection pq becomes the tangent to the projection of the curve at the projection of P.

9. The object of the method of projections is to extend our knowledge of curves from those which we already know, or can easily investigate, to others into which they project. Thus the properties of the ellipse can be deduced from those of the circle, and the rectangular hyperbola helps us to ascertain the properties of the hyperbola of unequal axes. The preceding propositions serve to connect the diameters and other lines, and the areas, of the known curve with those of its projection.

Examples.

1. If the inclination of two planes be the third of two right angles, and a line of 4 yards in one of the planes be inclined to the line of intersection at half the above angle, find the area of the square on the projected line.

2. In Art. 6 when D falls without AB, find the inclination of the planes in order that, in any given case, the angle ACB may not be altered by projection.

3. A triangle is projected on to a plane whose line of intersection with its own plane coincides with the base; it is projected back again on to its original plane, and the inclination of the plane is such that after these two projections its vertical angle is unaltered. Shew that this inclination gives the greatest possible increase to the vertical angle at the first projection.

4. Find the altitude of the sun when the greatest shadow of the side of a square on a horizontal plane equals its diagonal.

5. In what position must a cube be placed so that its shadow on a plane that receives the sun's rays directly may be the greatest possible?

CHAPTER II.

On the Cone and Sphere.

1. DEFINITIONS. If through O the centre of a circle ABC we draw a straight line OV at right angles to its plane, then the surface traced out by a straight line which passes through any point V of this line and the successive points of the circle ABC is called a *Cone;* or, more distinctively, a *Right Circular Cone,* OV is called the *Axis* of the cone, and V its *vertex.*

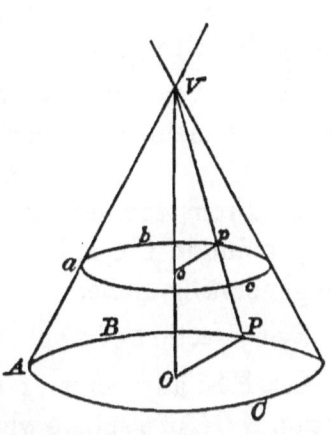

We repeat here Euclid's definition of a sphere:

A *sphere* is a solid figure described by the revolution of a semicircle about its diameter which remains fixed.

2. If VP be any position of the generating line of a cone, the angle OVP is the vertical angle of a right-angled triangle of which the height OV and the base OP are constant; therefore OVP is constant, and the angle between the axis and the generating line in all its positions is invariable. Also if any plane at right angles to the axis cuts the cone in the curve abc, abc will be a circle. Let o and p be the points

in which the plane cuts the axis and the generating line VP; join op.

Then op is the base of a right-angled triangle of which the height oV and the vertical angle oVp are invariable for all positions of VP; hence op is constant, and the section abc is a circle, with centre o.

Every plane containing the axis cuts the cone in two right lines inclined to each other at a constant angle.

3. From the definition of a sphere it is evident that every point in the sphere is equidistant from the centre of the generating circle.

Every section of a sphere by a plane is a circle.

Let P be any point of the section ABC of a sphere whose centre is O made by a plane. Draw ON at right angles to the cutting plane, and join OP, NP. Then for every point in the section the height ON and the hypotenuse of the right-angled triangle ONP are constant, and therefore the base NP is constant, and ABC is a circle of which N is the centre.

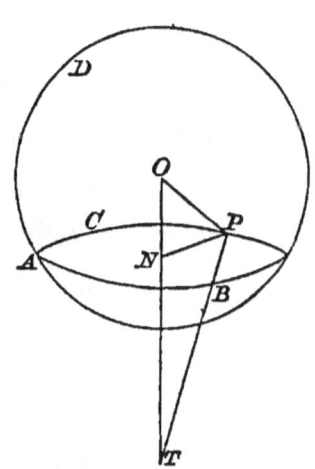

4. DEFINITION. A straight line is said to *touch* a *sphere* when it touches the circle in which the plane through the line and the centre of the sphere cuts the sphere.

It is evident that if a plane be drawn at right angles to the diameter of a sphere through its extremity, every line drawn in this plane through the extremity of the diameter will touch the sphere, and therefore the plane is said to touch the sphere.

5. All the lines drawn from one point to touch a sphere are equal.

Let TP (fig. § 3) be any line drawn from the point T to touch the sphere $ABCD$ whose centre is O in the point P. Join OP. Then TP touches a circle of which OP is a radius, therefore OPT is a right angle. Join OT. Then TP is the base of a right-angled triangle of which the height OP and the hypotenuse OT are constant.

Therefore all the lines drawn from T to touch the sphere are equal. Also the angle OTP is constant, and the equal tangents will therefore generate a cone of which T is the vertex and OT the axis. Also the points common to the sphere and cone lie on a circle whose plane is at right angles to OT.

CHAPTER III.

On the Ellipse.

1. DEFINITION. Let AVA' be a cone whose axis VO is in the plane of the paper, and VA, VA' its generating lines in that plane. And let the cone be cut by a plane APA' at right angles to the plane of the paper and intersecting it in the line AA' such that A, A' are points in the generating line on opposite sides of the axis VO; then the section APA' is called an *Ellipse*.

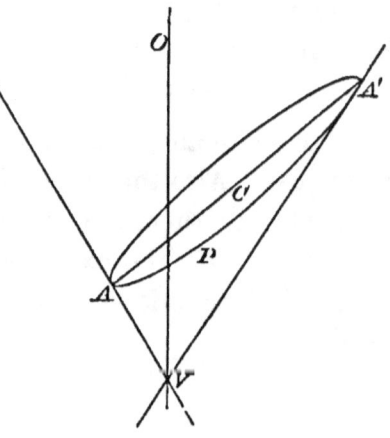

It is manifest that the ellipse will be a closed curve, and divided into equal parts by the line AA': every line in the plane of the section drawn at right angles to AA' will meet the curve in two points on opposite sides of AA', and at equal distances from it.

AA' is called the axis major of the ellipse, and a line through C its middle point in the plane of the curve at right angles to AA' limited by the surface of the cone is called the axis minor.

ON THE ELLIPSE.

2. Every ellipse may be considered as the projection of a circle whose diameter equals the axis major of the ellipse.

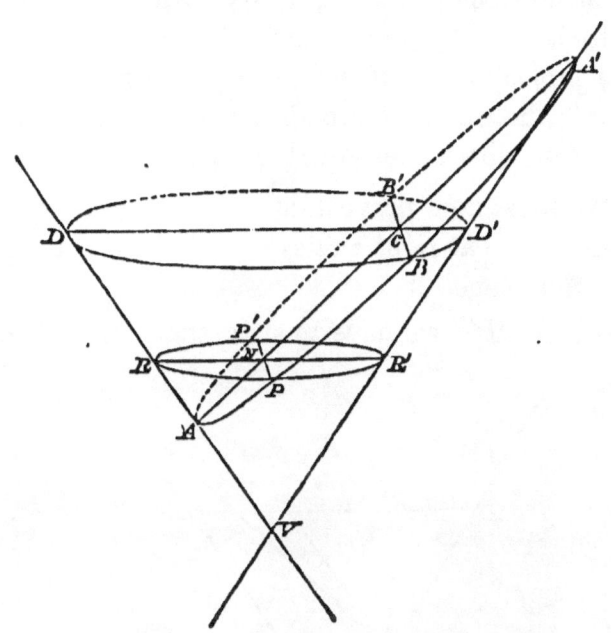

Let DBD', RPR' be two circular sections of the cone made by planes at right angles to the axis through C, N, the middle and any other point of AA', and cutting the plane of the curve in BCB', PNP' at right angles to AA'.

Then $NP^2 = RN . NR'$ and $BC^2 = DC . CD'$.
Also $RN : DC :: AN : AC$ and $NR' : CD' :: NA' : A'C$;
$\therefore RN . NR' : DC . CD' :: AN . NA' : AC^2$,
$\therefore NP^2 : BC^2 :: AN . NA' : AC^2$.

Now if NQ be the ordinate of a circle drawn on AA' as diameter, we shall have $NQ^2 = AN . NA'$,
and $\qquad NP^2 : BC^2 :: NQ^2 : AC^2$,
or $\qquad NP : BC :: NQ : AC$,
$\qquad NP : NQ :: BC : AC$;

then if a circle, diameter AA', be inclined to the plane of the paper at such an angle that the maximum ratio of diminution of the planes be $AC:BC$, every ordinate of the circle will be diminished by projection in this ratio, and will therefore after projection equal the corresponding ordinate of the ellipse, and the circle whose diameter is AA' will project into an ellipse whose axes equal AA', BB'.

3. We must now prove that in every case the diameter BB' of the ellipse at right angles to the plane of the paper is $< AA'$ in this plane.

$BC^2 = DC \cdot CD'$: we have to show that $DC \cdot CD' < AC^2$.

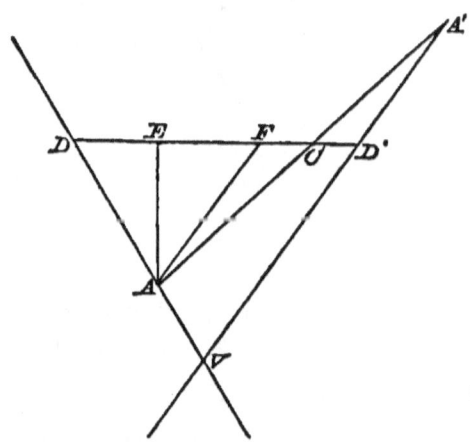

Draw AE perpendicular to DD', and AF parallel to $D'A'$ so that CF will $= CD'$.

Then $CA^2 = CE^2 + AE^2$
$= DC \cdot CF + EF^2 + AE^2$ (Euclid II. 6)
$= DC \cdot CD' + AF^2$ (or AD'^2),

thus $DC \cdot CD'$ is always $< CA^2$, and the ratio $BC : AC$ one of less inequality.

We may now trace some of the properties of the circle in the ellipse by means of the principles of projection already proved.

ON THE ELLIPSE. 13

4. In the circle if two diameters are at right angles to one another each bisects all chords parallel to the other. But by Chap. I. § 4, the projection of a point which bisects a line bisects the projection of the line; therefore chords of the circle which are bisected by a diameter will project into chords of an ellipse bisected by the projection of the dia-

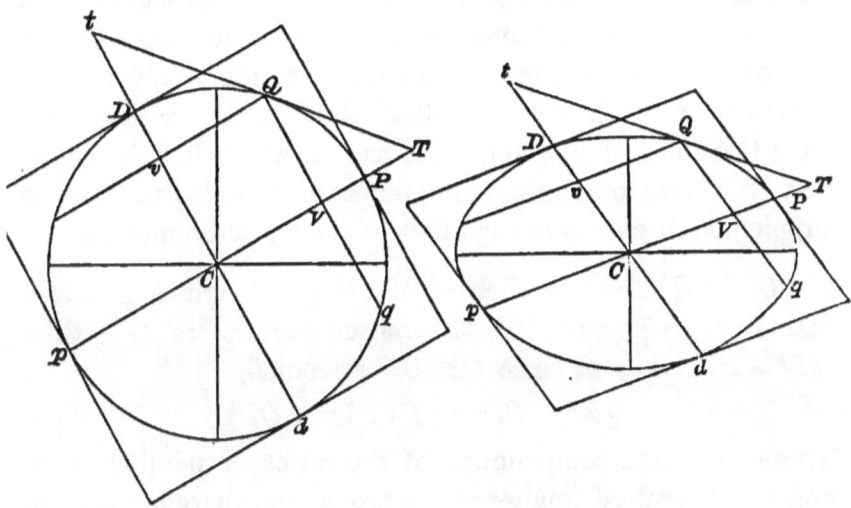

meter. Hence diameters of the circle at right angles to one another project into diameters of the ellipse which bisect each the chords parallel to the other. These diameters of the ellipse are said to be conjugate on account of the relation between them being mutual.

Also the tangents to a circle at the extremities of a diameter are at right angles to the diameter and parallel to the chords bisected by it: and the parallelism of straight lines is not destroyed by projection; hence the tangents at the two extremities of a diameter of an ellipse are parallel to the chords bisected by it and to the conjugate diameter.

Those angles at the centre contained by conjugate diameters in which the axis major lies will in all cases be less than a right angle.

The accompanying figure shews how diameters of the circle at right angles to one another project into conjugate diameters of the ellipse.

5. *Supplemental Chords.* The angle in a semicircle is a right angle in whatever point of the circumference the chords containing it meet: therefore when projected these chords are parallel to conjugate diameters of the ellipse: such chords when projected are called *supplemental chords;* they form two sides of a triangle the base of which is a diameter, and its vertex on the circumference of the ellipse. Hence supplemental chords are parallel to conjugate diameters.

6. If QV be a semi-chord of a circle at right angles to a diameter PCp, and CD the radius parallel to QV, then $QV^2 = PV \cdot Vp$; or since CP, CD are equal,
$$QV^2 : CD^2 :: PV \cdot Vp : CP^2.$$
These ratios are compounded of the ratios of parallel lines, and are therefore unaltered by projection, therefore also in the ellipse
$$QV^2 : CD^2 :: PV \cdot Vp : CP^2.$$

DEFINITION. The half of a chord bisected by a diameter (as QV by PCp) is called an *ordinate* to that diameter.

7. If QT, the tangent to the circle at Q, meet CP produced in T, $CV \cdot CT = CQ^2$ or $= CP^2$; or CP is a mean proportional between V and T. The ratios of these lines are not altered by projection; therefore also in the ellipse
$$CV \cdot CT = CP^2.$$

As a particular case of this proposition, if the tangent at P meets the axis major in T, and PN be the ordinate,
$$CN \cdot CT = CA^2.$$

8. The tangents to a circle at the extremities of any chord meet in the diameter (produced) which bisects the chord: and the tangent at the extremity of the diameter when terminated by the pair of tangents is bisected at its point of contact. These properties are not affected by projection, and therefore hold true in the ellipse.

COR. Hence to draw tangents to an ellipse from any point T, join CT and let it cut the curve in P; draw the tangent at P by the latter part of the last Art. In CP take a point V such that CV is a third proportional to CT, CP. Through V draw an ordinate QVq parallel to the tangent at P: join TQ, Tq, then by the proposition of the last Art. these will be the tangents required.

9. The above propositions might equally have been made to take the form, if Qv be the ordinate to CD, and the tangent at Q meet CD in t,

$$Qv^2 : CP^2 :: Dv \cdot vd : CD^2,$$
and $\qquad Cv \cdot Ct = CD^2.$

As a particular case of the latter, if Pn be the ordinate from P to the axis minor, and the tangent at P meet the axis minor in t, $\qquad Cn \cdot Ct = CB^2.$

10. $PF \cdot PG = BC^2$, $PF \cdot Pg = AC^2$.

Let the normal at P (*the line through P at right angles to the tangent*) meet the axes and DCd the diameter conjugate to CP in G, g and F; and let the ordinates PN, Pn produced meet Dd in R, r. Then in the quadrilateral $GFRN$, the angles at F and N are right angles; and the quadrilateral may be inscribed in a circle.

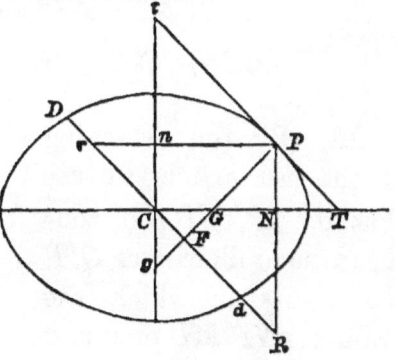

Hence $PG \cdot PF = PN \cdot PR = Cn \cdot Ct = BC^2$.

Similarly a circle on rg as diameter will pass through n and F, and $PF \cdot Pg = Pn \cdot Pr = CN \cdot CT = AC^2$.

11. $NG : NC :: BC^2 : AC^2$.

If we draw a circle on the axis major of the ellipse as diameter it is called the auxiliary circle, and will be equal to that which projects into the ellipse; and if the projected circle were made to revolve about its diameter AA', it would come to coincide with the auxiliary circle: and the point which, when the circle was projected, projected into P will now be found in the ordinate NP produced, and if this point be called Q we shall have $NP : NQ :: BC : AC$. Also the tangent to the circle which projects into the tangent PT passes through T, and when made to revolve about AA' will come into the position QT: hence the tangent to the auxiliary circle at Q will meet AA' produced in the same point as the tangent to the ellipse at P.

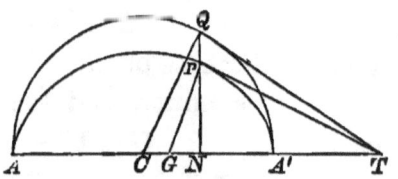

Hence we shall have

$NG \cdot NT = NP^2$ and $NC \cdot NT = NQ^2$;

$\therefore NG : NC :: NP^2 : NQ^2 :: BC^2 : AC^2$.

12. By the reasoning of the last article we see that if CP, CD be conjugate semi-diameters, Q, R the points in which the ordinates NP, MD produced meet the auxiliary circle, QCR will be a right angle.

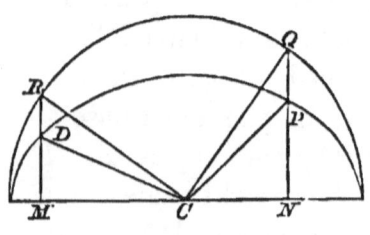

ON THE ELLIPSE.

Hence the triangles CQN, RCM will be equal in all respects. Hence

$CM = QN$, and $CN^2 + CM^2 = CN^2 + QN^2 = CQ^2 = AC^2$.

Also $QN = CM$, $RM = CN$; $\therefore QN^2 + RM^2 = AC^2$, and $PN^2 + DM^2 : QN^2 + RM^2 :: BC^2 : AC^2$; $\therefore PN^2 + DM^2 = BC^2$, hence $CP^2 + CD^2 = CN^2 + CM^2 + PN^2 + DM^2 = AC^2 + BC^2$.

13. In the circle if the chords Qq, Rr intersect in O, we have $QO \cdot Oq = RO \cdot Or$; and if we draw radii CP, CD parallel to the chords we have

$$QO \cdot Oq : CP^2 :: RO \cdot Or : CD^2.$$

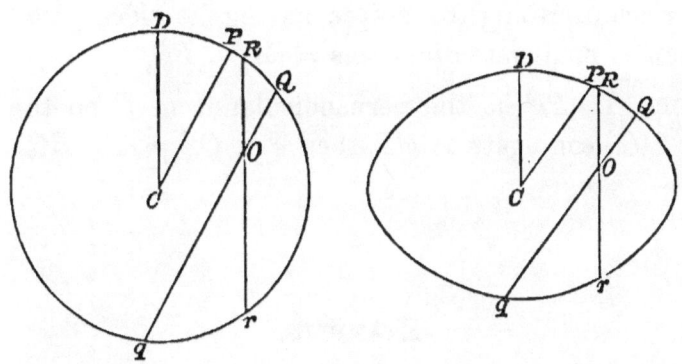

These ratios are not affected by projection, and the same proportion holds in the ellipse: it is independent of the position of O, but depends only on the direction in which the chords are drawn. It is equally true if O be a point exterior to the ellipse. And if tangents be drawn to an ellipse from any point, their lengths will be proportional to the parallel semi-diameters.

Hence if a circle intersect an ellipse in four points and the common chords be drawn, it will easily appear that the semi-diameters to which these chords are parallel must be equal, and therefore they and the chords parallel to them equally inclined to the axes.

14. The area of the parallelogram circumscribing an ellipse and touching it at the extremities of conjugate diameters $= 4AC \cdot BC$.

We have seen that the tangents at the extremities of any diameter are parallel, and if these be drawn at the extremities of two conjugate diameters they will form a parallelogram which is the projection of a square circumscribing the circle whose diameter $= 2AC$, and its area therefore $= 4AC^2$. But by projection this area is diminished in the ratio of $AC : BC$. Hence the area of the parallelogram circumscribing the ellipse having its sides parallel to any pair of conjugate diameters $= 4AC \cdot BC$.

Cor. If PF be the perpendicular from P on the diameter DCd conjugate to CP, then $PF \cdot CD = AC \cdot BC$.

Examples.

1. If an ellipse be projected on a plane intersecting its plane in BC, and having with its plane the maximum ratio of diminution $AC : BC$, it will project into a circle on BC as diameter.

2. If from any point P of an ellipse PQ be drawn parallel to the minor axis to meet the auxiliary circle in Q, and PR parallel to the major axis to meet the circle on the minor axis in R, QR produced will pass through the common centre.

3. A rectilinear figure circumscribing an ellipse and having its sides bisected at the points of contact is the projection of one regular polygon, and will project into another.

4. All such rectilinear figures of a given number of sides are equal in area however placed about the curve.

5. Two triangles circumscribing an ellipse have their sides bisected at the points of contact; the triangular corners cut off from the two triangles, from each by the sides of the other, are equal.

6. A rectilinear figure inscribed in an ellipse has one of its sides bisected at V, and CV produced meets the curve in P; if $CV : CP$ is a constant ratio the same for all the sides, the figure is the projection of one regular polygon and will project into another.

7. If PCP', DCD' are conjugate diameters, then PD, PD' are proportional to the diameters parallel to them.

8. All triangles circumscribing an ellipse which have the line joining each angular point with the point of contact of the opposite side passing through the centre, are equal and are less than any other that has not this property.

9. Any two diameters bisecting supplemental chords are conjugate.

10. Diameters which coincide with the diagonals of the parallelogram on the axes are equal and conjugate.

11. A straight line parallel to the axis major joins the extremities of equal conjugate diameters and intersects the circle on the axis minor in two points. The lines from the centre to these points are at right angles to one another.

12. The length of the line joining the extremities of equal conjugate diameters is independent of the magnitude of that axis to which it is at right angles.

13. The diagonals of any parallelogram formed by tangents at the extremities of conjugate diameters coincide with conjugate diameters.

14. The angular points of these parallelograms lie on one ellipse concentric, similar and similarly situated with the original.

15. In a circle, if the tangents be drawn at the extremities of a chord which passes through a fixed point within the circle, the point of intersection of the pairs of tangents lies always on a straight line. This is called the polar of the point which is called the pole. If Q be the pole, the polar is at right angles to CQ produced, and if CQ produced meets the polar in T, then $CQ \cdot CT =$ square on the radius. Prove this, and transfer the property to the ellipse, shewing that if CQT meets the ellipse in P, $CQ \cdot CT = CP^2$.

16. If in each of two ellipses the two axes are in the same ratio to one another, shew that radii which are equally inclined to either (say the major) axis in both bear to one another a constant ratio, i.e. the curves are similar.

17. Parallel elliptic sections of the same cone are similar.

18. If in Ex. 15, Q lies always upon an ellipse similar and similarly situated to the original ellipse, intersecting it and having its tangents at the points of intersection meeting in C, then T will be found upon the part of the second ellipse which lies without the curve.

19. The rectangle of the segments of any tangent intercepted between two parallel tangents made by the point of contact = square on the parallel semi-diameter.

20. The rectangle of the intercepts on parallel tangents made by any other tangent = square on the semi-diameter parallel to them.

21. If any tangent meet any two conjugate diameters, the rectangle under its segments = square on the parallel semi-diameter.

22. If from the extremity of each of any two semi-diameters ordinates be drawn to the other, the two triangles so formed will be equal in area.

23. Or if tangents be drawn from the extremity of each to meet the other produced, the two triangles so formed will be equal in area.

24. If CA, CB be the semi-axes of an ellipse and the rectangle $BCAD$ is completed: shew that if an ellipse similar and similarly situated to the given ellipse be described about it, and any chord $DPQR$ be drawn cutting the first ellipse in P, R, and the second in Q, $PQ = QR$.

25. Inscribe in a triangle an ellipse with axes parallel to given lines, and in a given ratio.

26. If the elliptic area between the two radii CP, CQ is invariable, prove that the area between the chord PQ and the curve will be so also.

27. PD joins the extremities of conjugate diameters, and CIJ makes a constant angle with PD, meeting it in I, and the parallel tangent in J. Shew that I and J trace similar curves.

28. If P be any point of an ellipse, and AP, $A'P$ produced meet the tangents at A', A in R and S, the tangent at P will bisect AS and $A'R$.

29. If from an external point O, two straight lines OAP, OQA' be drawn through the vertices of an ellipse $APA'Q$: if QA, $A'P$ intersect in R, OR is at right angles to the axis major.

30. If TP, TP' be tangents to an ellipse, and PCp be the diameter through P, then $P'p$ is parallel to CT.

31. If TP, TP' be tangents to an ellipse from an external point T, TR the diagonal of the parallelogram on TP, TP' and R be on the ellipse, then T will lie on an ellipse similar and similarly situated to the former.

32. If from any point T exterior to an ellipse, a line be drawn parallel to either axis to meet the curve the first or second

time in Q, the line bisecting TQ at right angles and that bisecting the tangents from T will meet on the tangent at Q.

33. Find the centre of a given ellipse.

34. Find the axes of a given ellipse.

35. In a given ellipse find the diameter conjugate to a given diameter.

36. If two points of a rod be constrained to move in two fixed lines which intersect at right angles, every other point of the rod will describe an ellipse.

37. If from any point P of an ellipse PQ be drawn to the axis major equal to BC, then if PQ produced either way meets the axis minor in R, $PR = AC$.

38. If PCP', DCD' are conjugate diameters, and PQ is drawn parallel to the axis major to meet the curve in Q; prove that DQ is parallel to two of the lines joining extremities of the axes of the curve.

39. If NP produced meets the auxiliary circle in Q, prove that GP, CQ produced meet on a circle whose diameter = sum of the axes of the curve.

40. Two ellipses have their axes equal each to each and in the same plane, also their centres coincident, draw the common tangents.

41. If CA, CB be any conjugate diameters of an ellipse and CB be produced to any point B', and an ellipse is described on CA, CB' as conjugate diameters; if the ordinate $P'PN$ be drawn parallel to BC, shew that the tangents to the ellipses at P, P' will intersect at a point lying on CA produced, and that

$$PN : P'N :: BC : B'C.$$

EXAMPLES.

42. If two ellipses with major axes parallel or at right angles intersect in four points, the opposite sides of the quadrilateral formed by joining the four points will be equally inclined to either axis of either curve.

43. If any system of diameters of an ellipse of a given even number divide it into equal sectors, the sum of the squares on the diameters is the same whatever their directions.

44. The same when the number is odd or even.

45. Prove that $PG \cdot Pg = CD^2$.

46. If CR be the perpendicular from the centre on the tangent at P, and BR, AD be joined; prove that the triangles ACD, RCB are similar.

CHAPTER IV.

Focal Properties of the Ellipse.

1. WE have now to shew that there are two points within the ellipse which, regarding the curve in its relation to the circle, are as it were a divided centre, and also two lines exterior to it which bear a remarkable relation to these points and the curve.

2. Let us premise the following propositions:

If the circle inscribed in the triangle VAA' touches the sides at K, S, L; and the circle that touches AA' and VA, VA' produced touches them at H, K', L', then $KK' = AA'$, and $AS = A'H$.

For the perimeter of triangle VAA'
$$= VA + AH + HA' + A'V$$
$$= VA + AK' + L'A' + VA' = VK' + VL' = 2VK'$$
$$= 2VK + 2KK':$$
also
$$= VK + KA + AA' + A'L + LV$$
$$= VK + AS + AA' + A'S + LV = 2VK + 2AA',$$
$$\therefore KK' = AA'.$$

Also $KK' = KA + AK' = AS + AH,$
$$\therefore AA' = AH + AS: \text{ but } AA' = AH + A'H,$$
$$\therefore AS = A'H.$$

FOCAL PROPERTIES OF THE ELLIPSE.

Also if we produce AA' both ways to meet LK, $K'L'$ produced in X, X', AX will $= A'X'$.

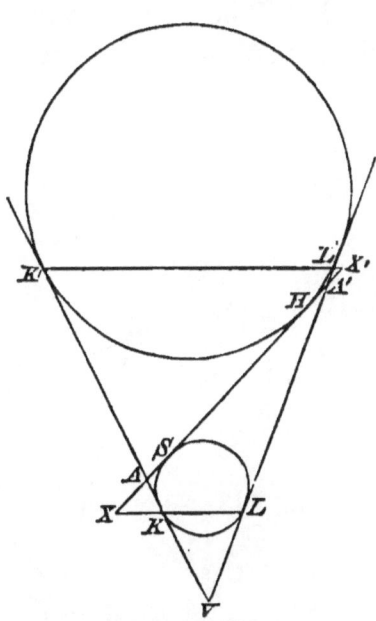

For AK, $A'L'$ being equal to AS, $A'H$ are equal, and being equally inclined to the axis of the cone will have equal projections on it. And the projections of AX, $A'X'$ will equal those of AK, $A'L'$, and are therefore equal: therefore AX, $A'X'$, being parts of the same line with equal projections on the axis of the cone, are equal.

3. Now to return to the construction in Chap. III. § 1. Let AA' be the intersection of the plane of the paper with a plane at right angles to it that cuts the cone VAA' in the ellipse APA'.

Inscribe in the triangle VAA' the circle SKL with centre O on the axis of the cone, and escribe the circle $HK'L'$ with centre O' also on the axis. Then if we make the circles to revolve about their diameters which coincide with the axis of the cone, they will generate spheres which will touch the cone in the circles KRL, $K'R'L'$. Every point of each of these circles is equidistant from V, and therefore the distance RR' from one circle to another along a generating line of the cone will be invariable. Also OS, $O'H$ will be at right angles to the cutting plane, which will therefore touch the spheres in S and H.

Let P be any point of the elliptic section, draw $VRPR'$ the generating line of the cone through P, and join SP, HP.

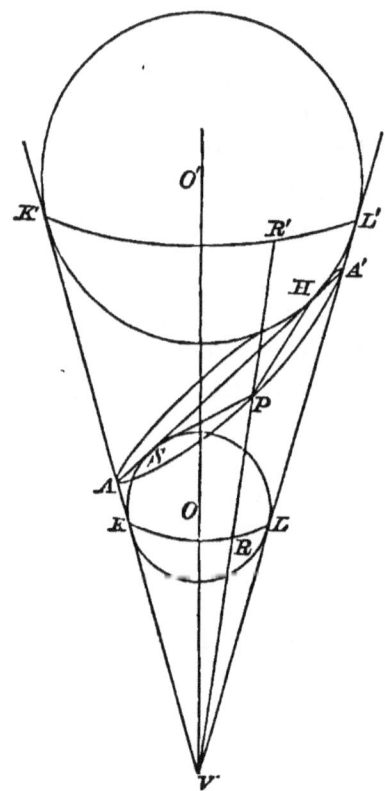

Then PS, PR are drawn from P to touch the sphere centre O in S and R, therefore $PS = PR$, and PH, PR' touch the sphere centre O' in H and R', therefore $PH = PR'$. Hence $SP + HP = PR + PR' = RR' = KK' = AA'$.

Hence the sum of the distances of any point of the ellipse from the points S, H within it is invariable and $= AA'$.

Hence we obtain the following construction to enable

FOCAL PROPERTIES OF THE ELLIPSE.

us to describe an ellipse: Fasten the two ends of a thread to the two points S and H, and let the thread be longer than SH: then stretch it with the point of a pencil, and mark the line which is traced by moving the pencil on all sides of the points so as to keep the thread tightly stretched: the curve so traced will be an ellipse. Viewing the ellipse in its relation to the circle, S and H may be considered as a divided centre, the sum of the distances of all points in the circumference from them being the same.

4. Now let the plane through P, perpendicular to the axis of the cone, cut the cone in the circle QPQ', and the plane of the ellipse in NP at right angles to AA'.

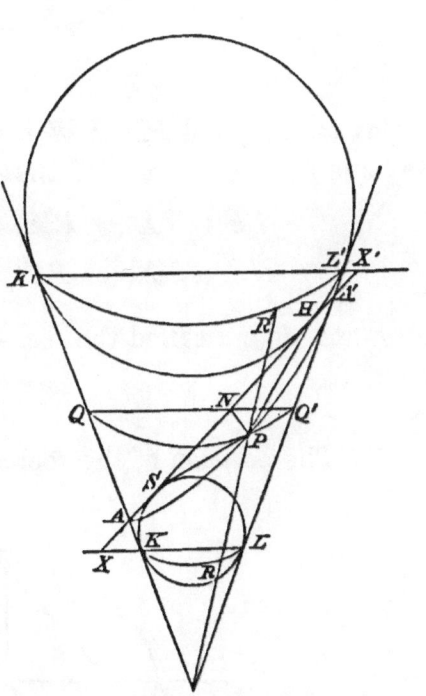

Then we shall have
$$SP = PR = QK,$$
and the triangles QAN, KAX are similar;
$$\therefore QK : NX :: AK : AX$$
$$:: AS : AX.$$

Hence whenever P is situated on the ellipse, $SP : NX$ is a constant ratio $= AS : AX$.

And similarly, joining HP,
$$HP (= PR' = Q'L') : NX' :: A'H : A'X'.$$

The ratios $AS : AX$, $A'H : A'X'$ are equal, and since $A'L' < A'X'$, each is a ratio of lesser inequality: either of them is called the *eccentricity*.

So then if we draw the elliptic section in the plane of the paper, and draw through X, X' lines XZ, $X'Z'$ at right

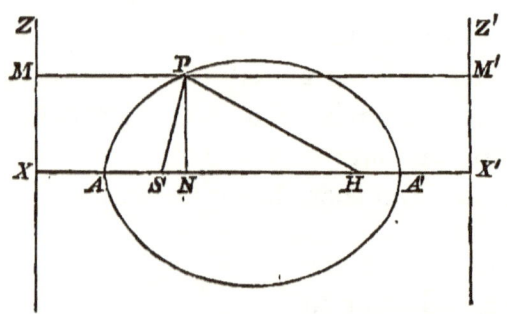

angles to XX', and PM, PM' perpendiculars to these lines; we have for any point P of the ellipse

$$SP : NX \text{ (or } PM) :: AS : AX,$$
$$HP : NX \text{ (or } PM') :: A'H : A'X'.$$

S and H are called the *foci* of the ellipse, XZ, $X'Z'$ the *directrices*.

5. The position of the foci and directrices is determined

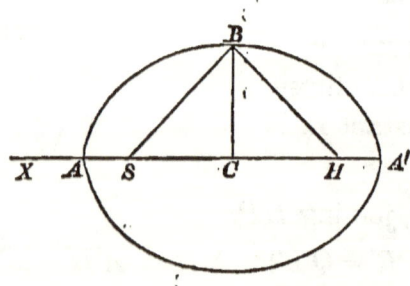

by the following relations: $CS^2 + CB^2 = CA^2$;

FOCAL PROPERTIES OF THE ELLIPSE. 29

For joining the foci with the extremity of the axis minor, $SB = HB = \frac{1}{2}(SB + HB) = AC$, and $SC^2 + BC^2 = SB^2 = AC^2$.

Also CX is a third proportional to CS and CA.

For first, $SB : CX$, i.e. $CA : CX :: SA : AX$.

Hence
$$CA : CX :: SA : AX,$$
$$CA : SA :: CX : AX,$$
$$CA - SA : SA :: CX - AX : AX,$$
or
$$CS : SA :: CA : AX,$$
$$\therefore CS : CA :: SA : AX,$$
$$:: CA : CX,$$

or CX is a third proportional to CS and CA.

6. Properties of the secant and tangent.

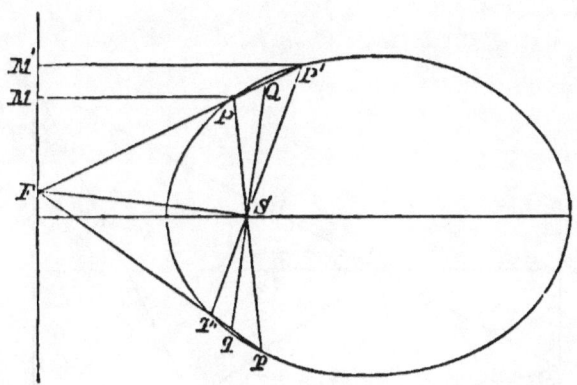

Let PSp, $P'Sp'$ be two focal chords. Let the secant $P'P$ cut the directrix in F, join SF. Let QSq bisect the vertical opposite angles PSP', pSp': PM, $P'M'$ perpendicular to the directrix.

Then $SP : SP' :: PM : P'M' :: PF : P'F$,

therefore SF bisects the angle PSp' (Euclid VI. A) and is perpendicular to QSq; hence also the secant pp' will pass through F.

Now let the secant FPP' revolve about F, so that P, P' will approach one another; SQ still bisecting the angle PSP' will be perpendicular to SF and therefore constant in position; and P, P' will finally coincide with Q which will then be a point on the curve, and the secant will then become a tangent to the curve at that point. Similarly if the secant $Fp'p$ turn about F, it will in its limiting position touch the curve in Sq produced. Hence the tangents at the extremity of a focal chord intersect on the directrix: and *the part of any tangent between the curve and the directrix subtends a right angle at the focus.*

7. The focal distances make equal angles with the tangent at any point.

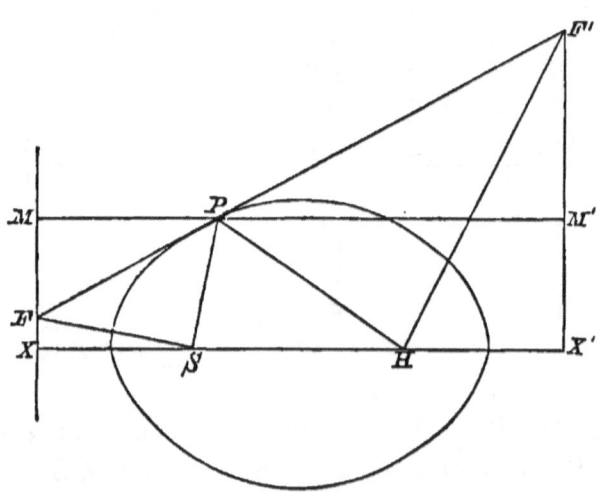

Let the tangent at P meet the directrices in F, F', join SF, HF'; PSF, PHF' will be right angles.

FOCAL PROPERTIES OF THE ELLIPSE.

Also the triangles MPF, $M'PF'$ are similar,

and $\quad SP : PM :: AS : AX :: HP : PM'$;

$\therefore SP : PH :: PM : PM' :: PF : PF'$,

or $\quad SP : PF :: HP : PF'$.

Hence (Euclid VI. 7) SPF, HPF' are similar triangles, and the angles SPF, HPF' are equal.

COR. The tangents at the extremities of the axes are at right angles to them.

8. The feet of the perpendiculars from the foci on any tangent lie on the auxiliary circle.

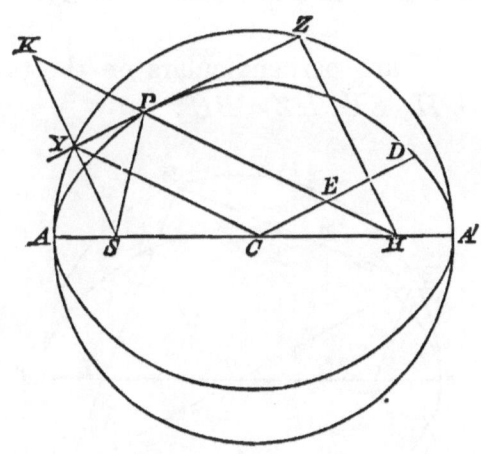

Let SY the perpendicular from S on the tangent at P meet HP produced in K. Join SP, CY. Then in the right-angled triangles SYP, KYP, PY is common and the angle SPY = the angle HPZ (by the last proposition)

$\quad\quad$ = the vertical opposite angle KPY.

Hence $SY = KY$, and $SP = KP$:

$\therefore KH = KP + PH = SP + PH = 2AC$.

And C, Y being the middle points of SH, SK, CY is parallel to HK and half of it and therefore $= AC$.

Hence Y lies on the circle on AA' as diameter (the auxiliary circle). And similarly Z the foot of the perpendicular from H.

COR. 1. CY being parallel to HP and bisecting SH, also bisects SP. Hence SYP being a right angle, the circle on SP as diameter passes through Y, and has its centre on CY; hence it touches the auxiliary circle at Y.

COR. 2. If CD be drawn parallel to the tangent at P and therefore conjugate to CP, and intersect PH in E, then $CEPY$ is a parallelogram, and $PE = CY = AC$.

9. If SY, HZ are perpendiculars on the tangent at P from the foci S, H; $SY \cdot HZ = BC^2$.

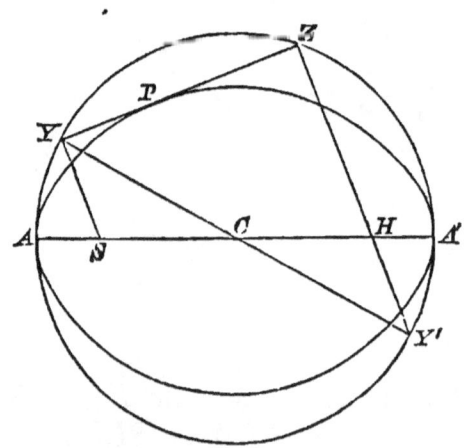

Since YZH is a right angle, YC produced will meet ZH produced on the auxiliary circle (at Y'). The triangles SCY, HCY' are equal in all respects, and

$$SY \cdot HZ = HY' \cdot HZ = AH \cdot A'H = A'C^2 - CH^2 = BC^2.$$

FOCAL PROPERTIES OF THE ELLIPSE.

10. $SP \cdot HP = CD^2$.

If we draw PF in fig. §8 perpendicular to DC (produced if necessary) we shall have the triangle PEF similar to HPZ and SPY; and $PE = AC$: hence

$$SP : SY :: HP : HZ :: PE : PF :: AC : PF$$

$$:: CD : BC, \because PF \cdot CD = AC \cdot BC;$$

$$\therefore SP \cdot HP : SY \cdot HZ :: CD^2 : BC^2, \text{ and } SY \cdot HZ = BC^2,$$

$$\therefore SP \cdot HP = CD^2.$$

11. If from any point Q of the tangent PK perpendiculars QN, QT be drawn to the directrix and SP, then

$$ST : QN :: AS : AX.$$

For QT is parallel to KS; and if we draw PM perpendicular to the directrix,

$$ST : SP :: QK : PK :: QN : PM,$$
$$\therefore ST : QN :: SP : PM :: AS : AX.$$

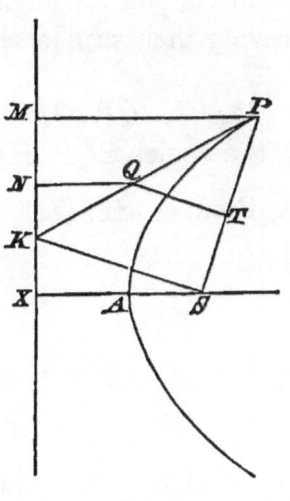

The student of analytical geometry will see in this proposition the basis of the polar equation to the tangent in terms of the angle PSA (α), viz. $\rho = \dfrac{a(1-e^2)}{\cos(\alpha-\theta) + e\cos\theta}$.

The proposition is due to Professor Adams, and the property is equally true of all the Conic Sections.

12. Hence if QP, QP' be the tangents to an ellipse from an exterior point Q, QP, QP' subtend equal angles at S.

For if we draw QT, QT', QM the perpendiculars on SP, SP' and the directrix, we shall have ST, ST' in the same proportion to QM and therefore equal.

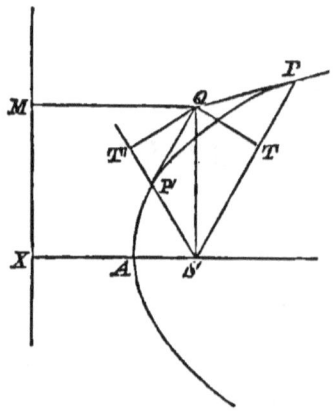

Hence the right-angled triangles QST, QST' are equal in all respects, and the angles QSP, QSP' equal.

If Q lie beyond the directrix, T, T' will lie in PS, $P'S$ produced, and the angles QSP, QSP' will be proved equal by proving their supplements equal.

13. If QP, QP' be the tangents to the ellipse from Q, the angles SQP, HQP' are equal.

Produce SP, HP' to R, R'; and let HP, SP' intersect in O.

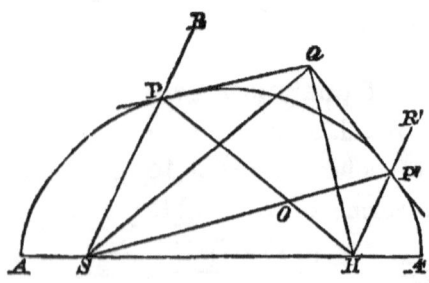

Then QP, QS bisect the angles HPR, $P'SP$ respectively, also $OPR = $ sum of the interior opposite angles OSP, SOP,

FOCAL PROPERTIES OF THE ELLIPSE. 35

and $\qquad QPR = $ sum of QSP, SQP.

$\therefore$ sum of OSP, $SOP = $ twice the sum of QSP, SQP,
of which $\qquad OSP = $ twice QSP,

$\therefore$ angle $SOP = $ twice SQP.

Similarly the angle $HOP' = $ twice HQP'.

And the angle $SOP = HOP'$, $\therefore SQP = HQP'$.

14. Any two tangents at right angles to one another intersect on a circle whose centre is C and square on the radius $= CA^2 + CB^2$.

Let any two tangents at right angles to one another cut

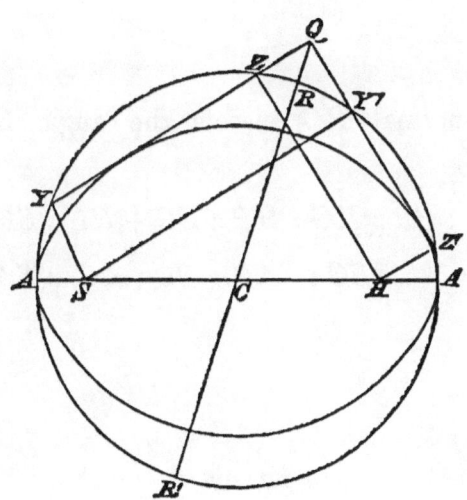

the auxiliary circle in Y, Z; Y', Z'. Draw SY, SY'; HZ, HZ'. Let QC cut the circle in R, R'.

Then SQ, HQ will be rectangles, with opposite sides equal.

3—2

Hence $QR \cdot QR' = QY' \cdot QZ' = SY \cdot HZ = BC^2$,

$$\therefore CQ^2 = CR^2 + QR \cdot QR'$$
$$= CA^2 + BC^2.$$

15. $\qquad SG : SP :: SA : AX.$

The tangent at P makes equal angles with SP, PH.

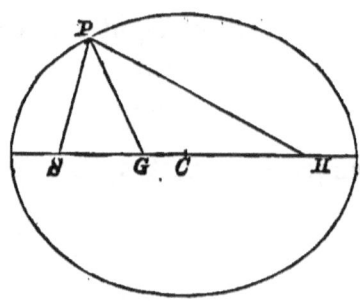

Hence the normal PG bisects the angle SPH. Hence (Euclid VI. 3)

$$SG : SP :: HG : HP :: SG + HG : SP + PH$$
$$:: 2SC : 2AC :: SC : AC :: SA : AX.$$

Examples.

1. A series of ellipses pass through a point and have a common focus and their axes major are equal: shew that the other focus always lies on the circumference of a fixed circle.

2. Under the same circumstances the centre also describes a circle.

EXAMPLES.

3. Under the same circumstances, what is the greatest excentricity the ellipse can have, and what does it then become?

4. Given three points of an ellipse and one focus, shew how to find the corresponding directrix.

5. The lines joining the extremities of a focal chord with the vertices have their points of intersection on the corresponding directrix.

6. The part of a directrix intercepted between the lines joining any point of the ellipse with the vertices subtends a right angle at the corresponding focus.

7. Two circles have their centres fixed and the sum of their radii constant: find the locus of the centre of a circle of constant radius that touches them both.

8. A circle through Y, Z, the feet of the perpendiculars on the tangent from the foci touches the axis major in Q, and has its centre O in the tangent: shew that $OQ = BC$.

9. If the tangent at a point P, the foot of whose ordinate is at N, intersects the major axis produced in T: prove that $TN \cdot TC = TA \cdot TA'$.

10. A circle through Y, Z, and N the foot of the ordinate at the point of contact, will also pass through the centre of the ellipse.

11. The focal distances of two points where the tangents are parallel form a parallelogram.

12. SY is perpendicular at the tangent at P, HY' parallel to PS meets YS produced in Y', prove $HY' = 2AC$.

13. If CF, perpendicular to the tangent at P intersect HP in F, $HF = AC$.

14. The locus of the intersections of the perpendicular from

the centre on the tangent with the focal distances is two circles with radius AC, and centres at the foci.

15. If CD cut SP and HP in E and E', prove $SE = HE'$ and the circle round SEC = that round $HE'C$.

16. If from the centre of an ellipse lines be drawn parallel and perpendicular to a tangent at any point, they inclose a part of one of the focal distances of that point which equals the other focal distance.

17. If SP, HP are at right angles to one another

$$SP \cdot HP = 2BC^2.$$

18. Given the two foci and one tangent of an ellipse to draw the directrices.

19. If from any point T of a tangent at P, TQ be drawn at right angles to SP produced if necessary and TRN at right angles to the axis major cuts the curve in R, then $SQ = SR$.

20. The tangent at the extremity of the latus rectum meets the axis major at the foot of the directrix.

21. The circle that passes through L, L' the extremities of the latus rectum through S and touches the nearer directrix is touched by LH.

22. Shew that lines drawn from a focus to points on the ellipse at equal distances from the extremities of the axis minor are equally inclined to the tangents at the points.

23. A ship sails over an elliptic path having its middle point at B, shew that she changes her direction as much as an observer at the focus does the direction of his telescope in watching her.

24. Given a focal chord and the tangents at its extremity, find the second focus.

EXAMPLES.

25. Given a focal chord and the second focus, find that which lies on the chord.

26. Given PO, QO tangents to an ellipse at the points P, Q; POQ is less than a right angle and it is known that one focus lies in PQ, find the other and the directrices.

27. If PN be an ordinate, the angle $PNY =$ the angle PSY.

28. If PN be an ordinate, PN bisects the angle YNZ.

29. If a line through X the foot of the directrix cut the ellipse in P, p, SP, Sp are equally inclined to either axis. If PSQ be a focal chord PS, SQ subtend equal angles at X.

30. The circle on PG as diameter cuts SP, HP in K and L, shew that KL is perpendicular to PG and bisected by it.

31. Prove $SG : HG :: SY : HZ$.

32. SZ and HY each bisect the normal PG.

33. If DR be the ordinate at D, and CD conjugate to CP, the triangles PGN, DRC are similar, and in any ellipse PG is proportional to CD.

34. If QPN be the common ordinate to the auxiliary circle and the ellipse, and the tangents at Q and P meet on the axis major in T, prove $TQ : TP :: BC : PG$.

35. If O, O' be the centres of the circles inscribed in the triangle SPH, and escribed on its side SH, then $PO . PO' = CD^2$.

36. If ET be the side of a parallelogram whose sides touch an ellipse at the extremities of conjugate diameters, and E, T when joined each with the two foci have their joining lines meeting in O and O'; then shew that O, S, O', H lie on the same circle; and that the sum of the angles subtended by SH at E, T and $O' =$ that it subtends at O.

37. A diameter CP produced intersects the directrix in V, prove that VS is at right angles to the diameter conjugate to CP.

38. Given the centre and directrix of an ellipse, also the directions of a pair of conjugate diameters, determine the position of the foci.

39. If tangents PT, QT meet on the auxiliary circle, prove that SP is parallel to HQ.

40. If a parallelogram circumscribes an ellipse, the lines joining the points of contact also form a parallelogram.

41. A parallelogram is described about an ellipse having two of its corners on directrices, prove that the other two will lie on the auxiliary circle.

42. If a line be drawn through a focus perpendicular to two parallel tangents, the rectangle of its segments made by the focus $= BC^2$.

43. A line is drawn through the focus of an ellipse perpendicular to a pair of parallel tangents: on this line as diameter a circle is described: prove that the chord of this circle parallel to the tangents such as when produced passes through the other focus $= BB'$.

44. If a quadrilateral be circumscribed about an ellipse, the angle subtended at either focus by opposite sides are supplementary.

45. Circles are described on SP, HP as diameters, and chords of these circles YI, ZJ are drawn at right angles to the axis major, prove that SI, HJ produced intersect on the axis minor.

46. The points in which the tangents at the vertices are intersected by any tangent are joined each with a focus; shew that these lines intersect in the normal.

47. The tangents at the extremities of the latera recta on the same side of the axis major intersect on the circumference of the circle through the foci and the points of contact.

EXAMPLES.

48. If tangents PT, QT meet in T and one of them QT is produced to any point Q', prove that the angle PTQ' is a mean between PSQ and PHQ.

49. An endless string of greater length than the circumference of an ellipse which is laid on a sheet of paper is made to pass round it and stretched tight by a pencil: prove that the point of the pencil will trace an ellipse having the same foci as the original. (It may be assumed that the normal to the pencil's path at each point will make equal angles with the directions of the string at the point.)

50. If P be the vertex of a triangle whose base AB is bisected in C. Then if $AP \cdot BP + CP^2$ is a constant quantity, the locus of P is an ellipse.

51. If AD be the portion of the generating line of a cone which contains the vertex of an ellipse cut from it intercepted between the vertex and a line through C at right angles to the axis of the cone; shew that $AD = CS$.

52. To cut an ellipse of given axes from a given cone.

CHAPTER V.

On the Hyperbola.

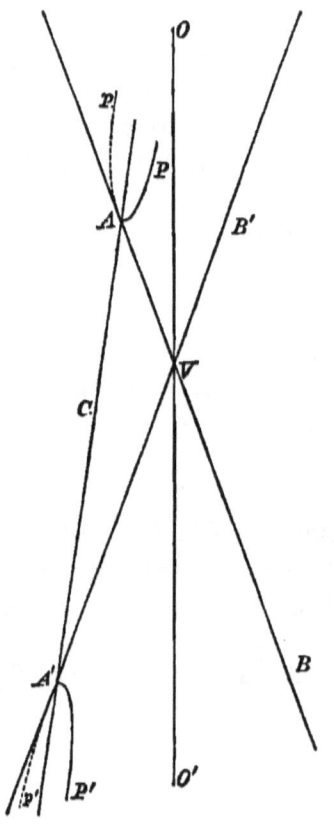

1. DEFINITION. Let AVB', BVA' be the two sheets of a cone, whose axis OVO' is in the plane of the paper and AB $A'B'$ the two positions of the generating line in that plane. Let the cone be cut by a plane APA' perpendicular to the plane of the paper which cuts it in the line AA', such that A, A' are points in the generating line of the cone on the same side of the axis OVO': then if this plane cuts the cone in the lines PAp $P'A'p'$, these lines make up a curve which is called an *Hyperbola*.

It is manifest that the two branches of this curve may be prolonged to any length and cannot intersect. Also they are each divided into two equal parts by the line AA' produced: every line in the plane of the section drawn at right angles to AA' will meet the curve in two points on opposite sides of AA' and at equal distances from it.

ON THE HYPERBOLA.

AA' is called the *transverse axis* of the hyperbola, and a line through C the middle point of AA', at right angles to AA' in the plane of the section and of a magnitude to be specified in the next article, is called its *conjugate axis*.

2. Every hyperbola may be projected from or into an hyperbola whose axes are equal, called an *equilateral* or *rectangular* hyperbola.

Let DED', RPR' be circular sections of the cone, made by planes at right angles to the axis through C and through

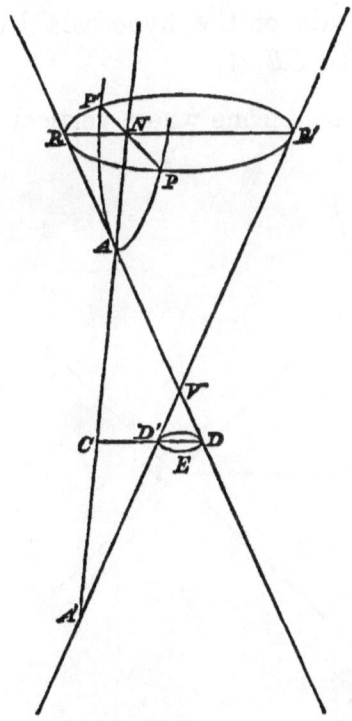

N any point of AA' produced, and let this latter plane cut the plane of the curve in PNP' at right angles to AA'. Then $NP^2 = RN \cdot NR'$.

Also in the similar triangles NAR, CAD

$$RN : DC :: AN : AC,$$

and in the similar triangles $NA'R'$, $CA'D'$

$$NR' : CD' :: A'N : A'C$$
$$:: A'N : AC$$
$$\therefore RN.NR' : CD.CD' :: AN.A'N : AC^2,$$

or $\qquad NP^2 : AN.A'N :: CD.CD' : AC^2.$

Now if we draw CE from C to touch the circle DED in E, $CE^2 = CD.CD'$: and we may now complete our definition of the conjugate axis of the hyperbola by saying that it is equal in length to CE.

Now let us take a cone whose vertical angle is a right

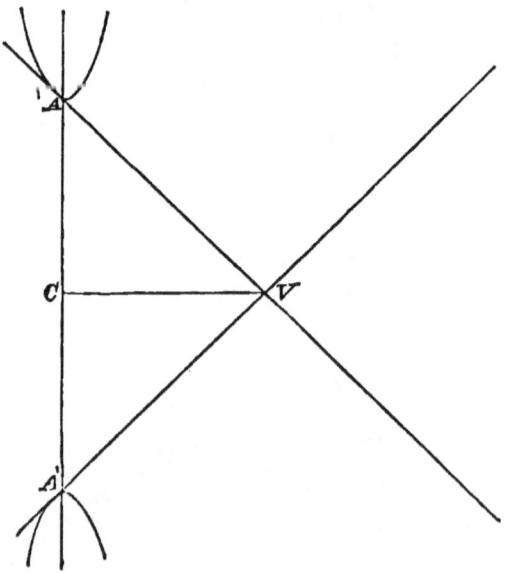

angle, and let an hyperbola be cut from it by a plane parallel to the axis and such that the transverse axis shall equal AA', then D, D' will merge in V, and CE will $= CV$, will also $= CA$, since the circle with centre C and distance CA or CA' will pass through V, AVA' being a right angle.

Hence the hyperbola will be rectangular and will have each of its axes equal to AA'.

Let the two hyperbolas be now placed in the same plane with their transverse axes coincident, and let the ordinate NP of the first hyperbola produced if necessary meet the rectangular hyperbola in Q. Then we have

$$NP^2 : AN . A'N :: CE^2 : AC^2,$$

and $\quad NQ^2 : AN : A'N :: AC^2 : AC^2$, or $NQ^2 = AN . A'N$:

hence $\quad NP^2 : NQ^2 :: CE^2 : AC,$

or $\quad NP : NQ :: CE : AC.$

We shall shew that CE may be greater or less than CA: if CE be greater than CA the first hyperbola will project into the rectangular one if, the transverse axes remaining coincident, they are placed in planes inclined to one another at the proper angle. If CE be less than CA the rectangular hyperbola will project into the one of unequal axes.

COR. The proposition proved above

$$NP^2 : AN . A'N :: CE^2 : AC^2$$

might have been equally proved if we had taken the circular section in the other sheet of the cone below A'. Hence we observe that if we take two points in the transverse axis of the curve produced both ways on opposite sides of C and equidistant from it, the ordinates drawn through them will be equal, and the points in which they meet the curve will be equidistant from the centre. Hence the two branches of the curve are equal and every line through the centre joining them is bisected in the centre.

3. It has been shewn that when the cone has a right angle at the vertex, and the cutting plane is parallel to its axis, the two axes of the curve are equal. Also if the cutting plane is parallel to the axis of the cone, $CE = CA >$ or $< CA$ according as angle of the cone is obtuse or acute. Hence it appears that the axes of an hyperbola may be equal or either in excess of the other. It may be shewn that when the angle of the cone is acute the transverse axis is the greater for all sections; when the angle is a right angle the axes are equal when the cutting plane is parallel to the axis of the cone, and the transverse axis the greater for all other sections. When the angle of the cone is obtuse, either axis may be greater than the other, or they may be equal according to the inclination of the cutting plane to the axis of the cone. This will appear if we draw VE at right angles to the axis of the cone to meet AA' in E and turn AA' about E, observing that
$$CD.CD' : AC^2 :: VE^2 : AE.EA',$$
and that $AE.EA'$ is least when AA' is at right angles to EV and diminishes the more the further it is turned from that position.

Further, it is evident that hyperbolas cut from any cone will have their axes in the same proportions when the cutting plane is so situated that $CD.CD' : AA'^2$ is a constant ratio. This will be the case when the plane moves parallel to itself: as in that case each of the ratios $CD : AC$ and $CD' : A'C$ remains constant.

If the angle of the cone is made to vary as well as the position of the cutting plane, a curve equal in all respects will be obtained if AA' remains unchanged and the rectangle $CD.CD'$ is also constant.

We may now proceed to investigate the properties of the rectangular hyperbola, and then we can generalise them by projection for the hyperbola of unequal axes.

4. Property of the Asymptotes.

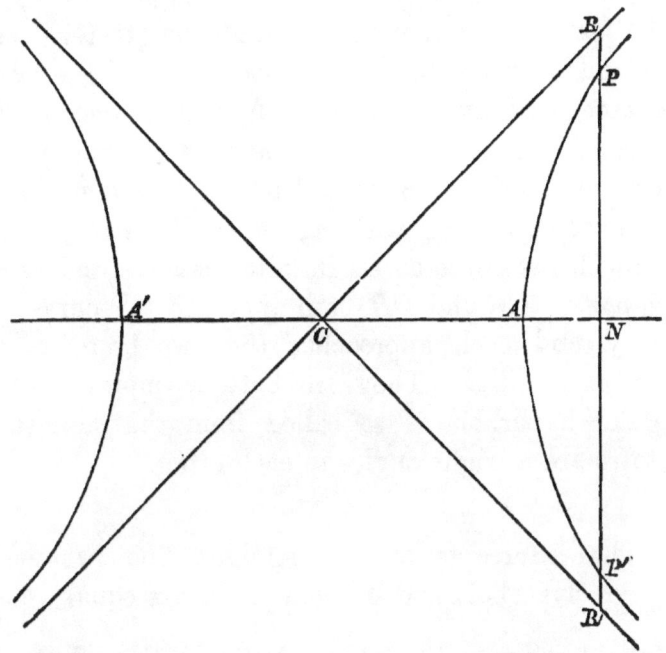

Let PAP' be one branch of the curve, and through C draw CR, CR' each making half a right angle with CA: let $RPNP'R'$ be at right angles to $A'A$ produced. Then $NR = NR' = CN$, and since $A'A$ is bisected in C, and produced to N,

$$\therefore AN \cdot A'N + AC^2 = CN^2:$$

but $$AN \cdot A'N = NP^2,$$

$$\therefore NP^2 + AC^2 = CN^2,$$

or $$NP^2 + AC^2 \cdot = NR^2.$$

But RR' is divided into two equal parts in N and unequally in P:

$$\therefore NP^2 + RP.PR' = NR^2.$$

$$\therefore RP.PR' = AC^2.$$

Now the further N is removed from A, the greater does RR' become, and therefore the greater does PR' and the less does PR become, and by increasing AN and therefore PR' we may make PR smaller than any assigned quantity; hence the curve approaches nearer and nearer to the straight line without ever coinciding with it; hence the straight line is said to touch the curve at an infinite distance, and is called an *asymptote*. CR and CR' both approach the curve in the same way, and each approaches the two branches at its opposite extremities. They are both asymptotes and the rectangular hyperbola is so called from the fact, that its asymptotes are at right angles to each other.

5. The intercepts on any straight line between the rectangular hyperbola and its asymptotes are equal.

In the last Art. since $NP = NP'$ and $NR = NR'$, therefore $PR = P'R'$, this proves the proposition for lines at right angles to the transverse axis.

Let any chord Qq be produced to meet the asymptotes in U and u. Let the ordinates at Q, q, meet the asymptotes in R, R', r, r'.

Then $\quad QU : QR :: qU : qr$,

$\quad\quad\quad Qu : QR' :: qu : qr'$.

Hence $\quad QU.Qu : QR.QR' :: qU.qu : qr.qr'$,

ON THE HYPERBOLA.

therefore $\quad QU \cdot Qu : AC^2 :: qU \cdot qu : AC^2,$

therefore $\quad QU \cdot Qu = qU \cdot qu.$

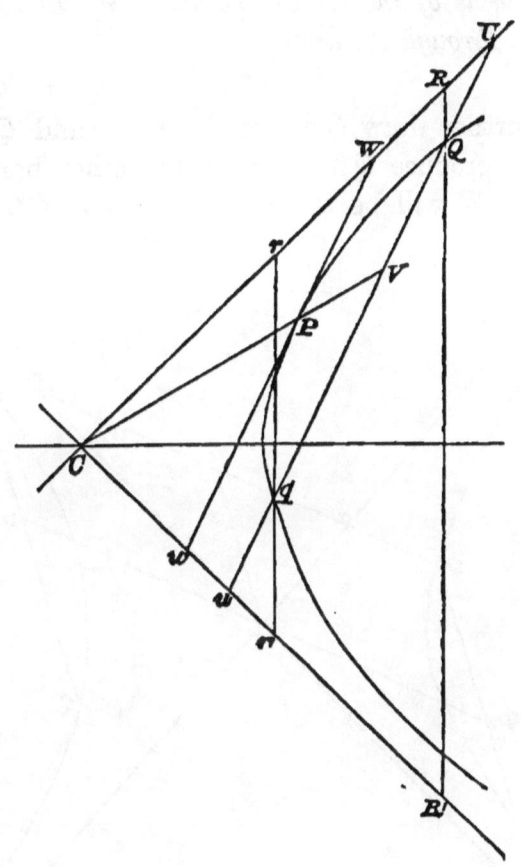

Let V be the middle point of Uu,

then $\quad QU \cdot Qu + VQ^2 = VU^2 = Vu^2 = qU \cdot qu + Vq^2,$

but $\quad QU \cdot Qu = qU \cdot qu,$ therefore $VQ^2 = Vq^2$;

hence V is also the middle point of Qq, and therefore $QU = qu.$

Join CV, cutting the curve in P: all chords of the asymptotes parallel to Qq will be bisected by CV, and therefore also all chords to the curve parallel to the same line. *The locus of the middle points of parallel chords is a straight line through the centre.*

If we further draw CW parallel to Qq and QW parallel to VC, and produce QW to meet the other branch of the curve in Q', W will be the middle point of QQ'.

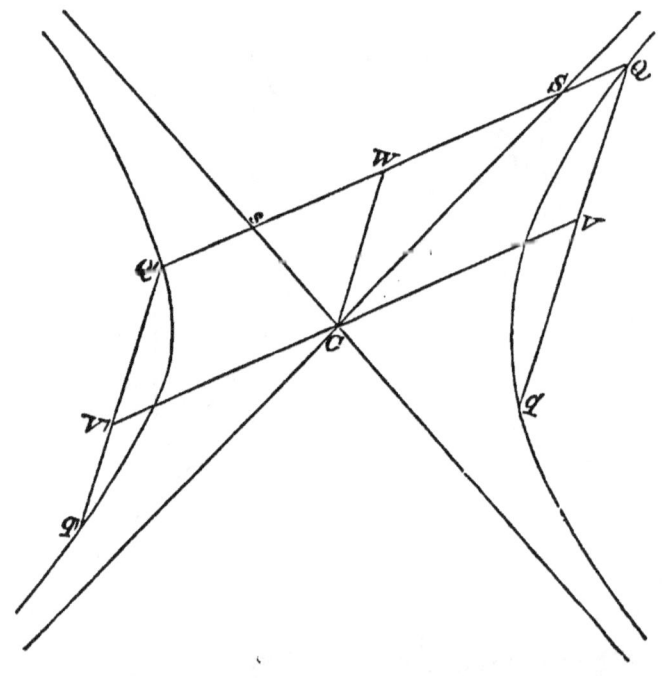

For the two branches of the curve are in all respects, equal and similarly situated with respect to the asymptotes. Therefore if we draw the chord $Q'q'$ parallel to Qq, VC produced will bisect $Q'q'$ (in V' suppose).

Also $Q'V'$ by construction $= WC = QV$:
hence $Q'q'$ will equal Qq, and CV' will equal CV:
$Q'C$, WV are parallelograms, and therefore
$$Q'W = V'C = VC = QW$$
and W is the bisection of QQ'.

And in like manner all chords between the two branches of the curves parallel to QQ' are bisected by the diameter CW.

Let QQ' meet the asymptotes in S, s: then since the angle at C is a right angle and V is the middle point of Uu, $CV = VU$, and CV, VU are equally inclined to the asymptote; so $\therefore$ are SW, WC being parallel to them:
$$\therefore SW = WC = Ws$$
since the angle at C is a right angle: W bisects Ss, and since it also bisects QQ', $\therefore QS = Q's$. Shewing that every straight line that cuts the curve and the asymptotes, each in two points, and the curve in one branch or both, has the intercepts between the curve and the asymptotes equal.

Cor. Hence as the chord Uu moves parallel to itself, so that Q and q approach each other and finally coincide in P, the chord ultimately becomes a tangent at P and is bisected in P: shewing that *the part of the tangent intercepted between the asymptotes is bisected at the point of contact.*

Each half of the tangent will equal CP.

6. $QV^2 + CP^2 = VU^2$.

Let Ww be the chord to the asymptotes which touches the curve at P. Through P draw rPr' the double ordinate to the asymptotes. Then
$$QU : QR :: PW : Pr :: CP : Pr$$

and
$$Qu : QR' :: Pw : Pr' :: CP : Pr';$$
$$\therefore QU.Qu : QR.QR' :: CP^2 : Pr.Pr'$$

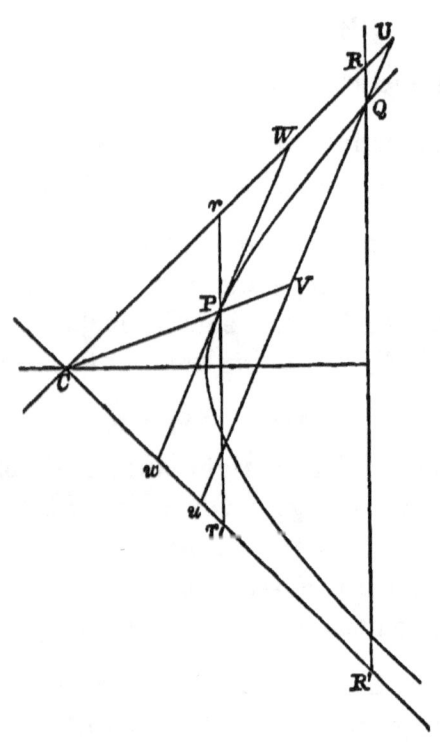

or
$$QU.Qu : AC^2 :: CP^2 : AC^2,$$
$$\therefore QU.Qu = CP^2 \text{ and } QV^2 + CP^2 = VU^2.$$

7. *The tangents at the extremities of any chord meet in the diameter which bisects it.*

Let Pp be any chord, Qq a parallel chord adjacent to it, then Vv through their middle points will pass through the centre of the curve.

Join PQ and let it meet the diameter in T.

Then

$$VT : vT :: PV : Qv :: pV : qv,$$

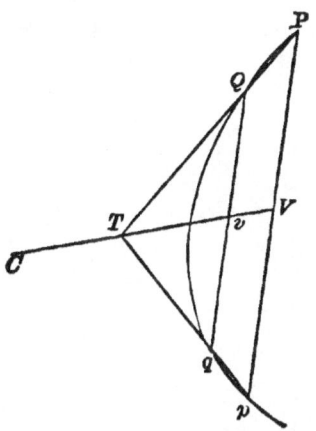

∴ the secant pq also passes through T. Thus at whatever distance Qq may be from Pp, PQ, pq always intersect on the diameter CV. Hence when Qq approaches Pp and finally coincides with it so that PQ, pq, become finally the tangents at P, p, these tangents will meet in the diameter CV.

8. $CV \cdot CT = CP^2$.

If QV be an ordinate of any diameter CP, and QT the tangent at Q, then CP is a mean proportional between CT and CV.

Let PR be the tangent at P intersecting QT in R, PR is parallel to QV: draw PO parallel to QT to meet QV in O.

Join OR; it will be the diagonal of the parallelogram $OPRQ$ and will therefore bisect the chord QP; hence (by the last proposition) RO bisecting the chord PQ and passing through the intersection of the tangents at its extremities is a diameter, and when produced will pass through C.

Hence $\qquad CV : CP :: CO : CR :: CP : CT$

and $\qquad CV \cdot CT = CP^2$.   Q. E. D.

Cor. 1. If PN be the ordinate from P to the transverse axis, and the tangent at P intersects the axis in T, we shall have $\qquad CN \cdot CT = CA^2$.

54   CONIC SECTIONS.

COR. 2. Hence to draw tangents to an hyperbola from any point T, join CT and produce it to meet the curve in P; draw the tangent at P by the aid of COR. 1.

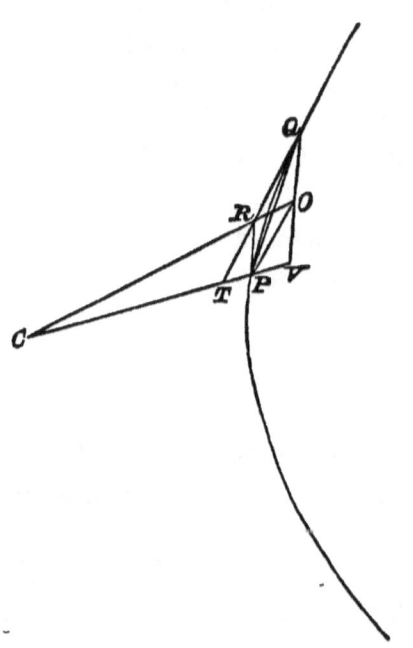

Produce CP to a point V such that CV is a third proportional to CT, CP: through V draw an ordinate QVq parallel to the tangent at P: join TQ, Tq; these (by the Proposition) will be the tangents required.

9. The Conjugate Hyperbola, Conjugate Diameters.

If on the conjugate axis of a rectangular hyperbola as transverse axis we draw another rectangular hyperbola, it is called the *conjugate hyperbola*, and it is manifest that it has the same asymptotes as the first.

ON THE HYPERBOLA. 55

Moreover if we draw semi-diameters CP, CD to meet the two curves equally inclined to the asymptotes and therefore to the axes, these will also be equal: and if we join PD it will cut the asymptotes at right angles, in O suppose. Take OR on the asymptote equal to CO and join PR, DR. Then $PR = CP$ and therefore touches the hyperbola at P; and DR ($= CD$) touches the conjugate at D. Also $CPRD$ is a parallelogram.

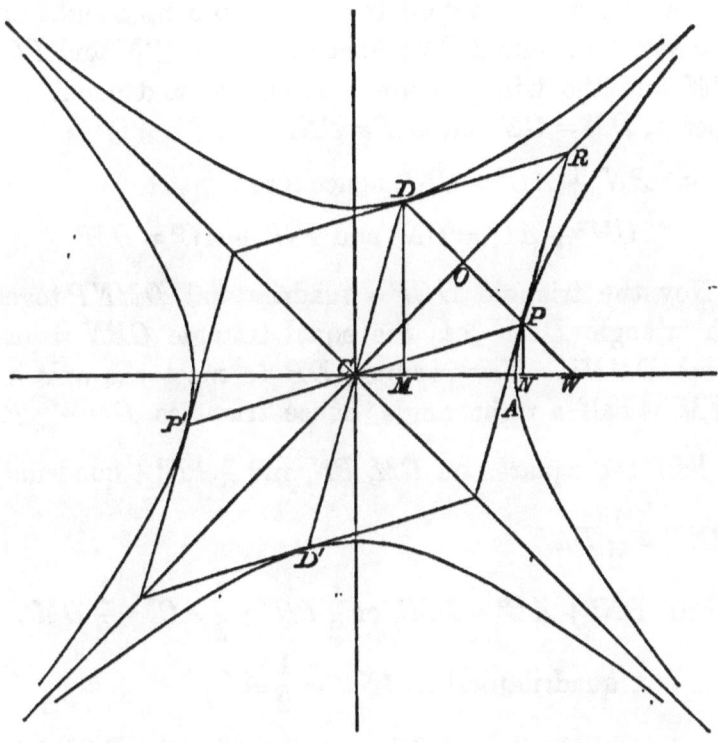

Moreover CP produced bisects all chords of the hyperbola parallel to PR or to CD, and similarly CD produced bisects all chords of the conjugate parallel to DR or to CP. Hence CP, CD produced each bisects the chords parallel to the other, and are called *conjugate* semi-diameters.

It will also be seen that CD will bisect the chords between the two branches of the curve parallel to PCP', and CP will bisect all chords of the conjugate parallel to DCD'.

10. The area of the parallelogram $CPRD$ is constant $= AC^2$.

Draw PN, DM at right angles to the transverse axis: the triangles CPN, CDM are equal in all respects. For the angles DCM, PCN are equally in excess and defect of half a right angle, and therefore together make up a right angle, as do also CPN and PCN; hence $DCM = CPN$ and $PCN = CDM$ and the triangles are equiangular and equal in all respects; $DM = CN$ and $CM = PN$.

But $PN^2 + AC^2 = CN^2$, hence also

$$CM^2 + AC^2 = CN^2 \text{ and } PN^2 + AC^2 = DM^2.$$

Now the triangle $DCP =$ quadrilateral $DMNP$ together with triangle DCM less the equal triangle $CPN =$ quadrilateral $DMNP$. Now produce DP to meet the axis in W, DWM is half a right angle, hence triangles DMW, PNW are half the squares on DM, PN, and $\frac{1}{2} PN^2 +$ quadrilateral $DMNP = \frac{1}{2} DM^2$.

But $PN^2 + AC^2 = DM^2$ or $\frac{1}{2} PN^2 + \frac{1}{2} AC^2 = \frac{1}{2} DM^2$,

$\therefore$ quadrilateral $DMNP = \frac{1}{2} AC^2$.

Hence parallelogram $DCPR =$ twice triangle $DCP =$ twice quadrilateral $DMNP = AC^2$.

Cor. 1. The parallelogram formed by tangents at the extremities of P, P' and D, D' of the diameters PCP', DCD' will have the constant area $4AC^2$.

ON THE HYPERBOLA. 57

COR. 2. The rectangle $PO \cdot OC =$ half the parallelogram $CPRD = \frac{1}{2} AC^2$.

COR. 3. If PF be the perpendicular from P on the conjugate semi-diameter CD, we shall have $PF \cdot CD = CA^2$.

11. $QV \cdot Vq$ is proportional to PV.

In the rectangular hyperbola, if a double ordinate Qq to the transverse axis cuts PV, a line from any point of the curve parallel to the asymptote CR in V, then we shall

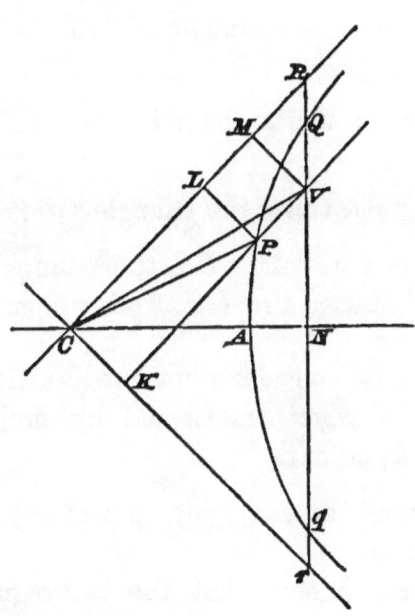

have $QV \cdot Vq$ proportional to PV. Join CP, CV, and we shall have $QV \cdot Vq = 4$ times the triangle CPV. For Qq is bisected in N,

$$\therefore QV \cdot Vq + VN^2 = QN^2,$$
$$\therefore QV \cdot Vq + VN^2 + AC^2 = QN^2 + AC^2 = RN^2.$$

but $\quad RV.Vr + VN^2 = RN^2,$

$\therefore QV.Vq + AC^2 = RV.Vr.$

Let VP produced cut Cr in K; draw VM, PL at right angles to CR: then $RV : Vr :: MV : Kr :: MV : VK,$

$\therefore RV.Vr : RV^2 :: MV.VK : MV^2,$

or $\quad RV.Vr : MV.VK :: RV^2 : MV^2:$

but $\quad RV^2 = 2MV^2$, $MV.VK =$ twice the triangle CKV;

$\therefore RV.Vr = 4$ times the triangle $CKV,$

and $AC^2 =$ twice the rectangle $PL.LC = 4$ times the triangle $CKP,$

$\therefore QV.Vq + 4$ times the triangle $CKP = 4$ times the triangle $CKV,$

$\therefore QV.Vq = 4$ times the triangle CPV:

and since PV is parallel to CR, the triangle CPV is proportional to PV: hence also $QV.Vq$ is proportional to PV.

12. Now let us consider what modifications these propositions undergo when transferred by projection to the hyperbola of unequal axes.

Let us suppose the conjugate axis of any hyperbola less than the transverse.

Then we have shewn that the rectangular hyperbola whose axes are equal to the transverse axis will project into it if the planes of the two hyperbolas intersect in AA' and are inclined at the proper angle. Also the asymptotes will project into lines equally inclined to the transverse axis and at such an angle that if DAD' is drawn to them through A at right angles to CA, AD or AD' will $= BC$ (the conjugate semi-axis).

ON THE HYPERBOLA.

Also the conjugate rectangular hyperbola when projected will have BC for its transverse, CA for its conjugate semi-diameter. This is the conjugate hyperbola: both the hyperbola and its conjugate will continually approach the asymptotes CD, CD' without ever meeting them.

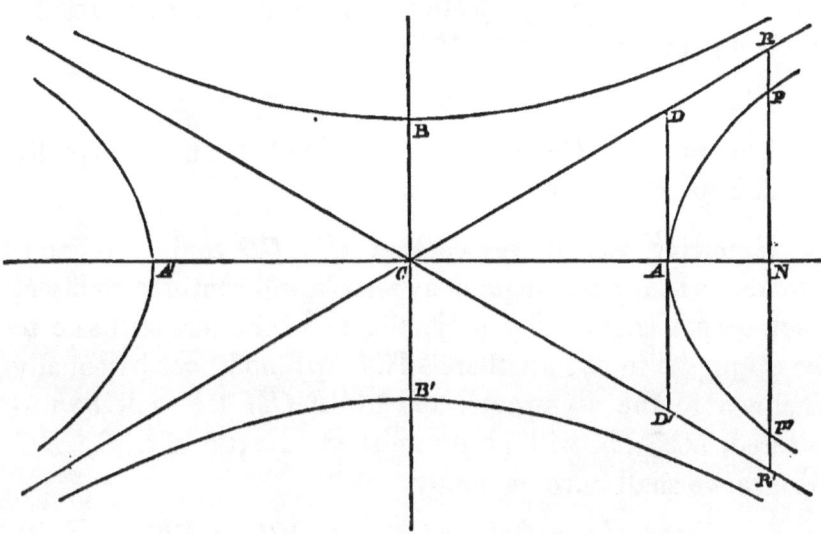

Also PR, $P'R'$, the distances between the curve and asymptotes on the double ordinate will be equal, and the tangents at A, A' will be at right angles to the transverse axis.

All lines perpendicular to the transverse axis will be diminished by projection in the ratio of $AC : BC$, hence $RP \cdot PR'$ which was equal to AC^2 will become BC^2.

Also V the middle point of any chord Uu of the asymptotes will also be the middle point of the chord of the curve as in § 5: and all chords to the curve parallel to Qq will be bisected by CV.

Also the tangent at P where CV cuts the curve will be parallel to QV and will be bisected at P: but as the angle WCw is now acute, each half of the tangent is less CP.

The proposition of § 6 must be modified by substituting the equal semi-conjugate diameter CD for CP; CD is parallel to Uu, and hence QV, CD, VU will be all altered in the same ratio by projection, and we shall have in the projected curve $\quad QV^2 + CD^2 = QU^2$

or $\qquad\qquad QU \cdot Qu = CD^2.$

The proof of the properties of §§ 7, 8 applies equally to all hyperbolas, rectangular or not.

Referring to § 9 we see that CP, CD conjugate semi-diameters of the rectangular hyperbola will continue to bisect each other's chords after projection: and they are still said to be conjugate to one another. They will no longer be equally inclined to the asymptote, and whilst CM, CN will be unaltered, DM, PN will be diminished in ratio of $AC : BC$. Hence we shall have generally

$$CM^2 + CA^2 = CN^2 \text{ and } PN^2 + CB^2 = DM^2,$$

and by addition

$$PN^2 + CN^2 + CB^2 = DM^2 + CM^2 + CA^2,$$

or $\qquad\qquad CP^2 + CB^2 = CD^2 + CA^2.$

Also the parallelogram $CPRD$ will be a parallelogram after projection, and its area will be diminished in the ratio of $AC:BC$, and $\therefore = AC.BC$: and the parallelogram formed by tangents at the extremities of conjugate diameters will $= 4AC \cdot BC$.

The proposition of § 11 will not be affected by projection, and therefore applies equally to the hyperbola of unequal axes, and it will be extended by a subsequent proposition

(Art. 15.) to the intercepts on a line drawn from any point in the curve parallel to an asymptote made by parallel lines in any direction.

13. If the conjugate axis of an hyperbola is greater than the transverse, we may project it into a rectangular hyperbola having its axes equal to the transverse, and the same results will obtain as in the last Article, except that the angles between the asymptotes that contain the transverse axis will be obtuse instead of acute.

14. Supplemental Chords.

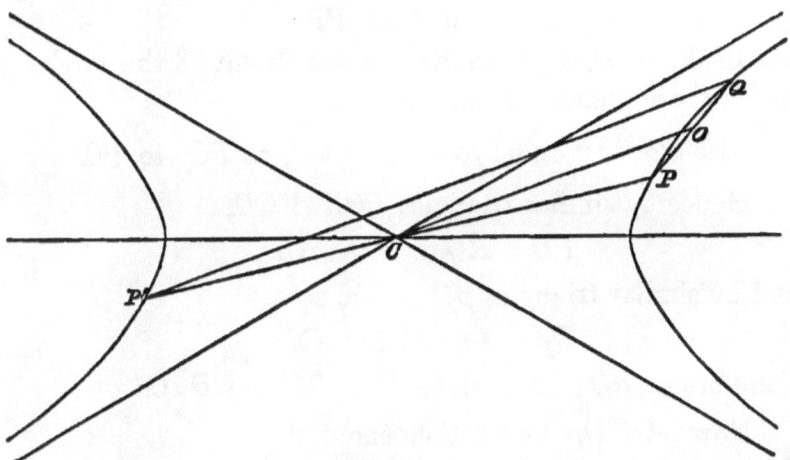

If any point Q in an hyperbola be joined with the extremities PP' of any diameter, PQ, $P'Q$ are called *supplemental chords*, and are parallel to conjugate diameters.

For if we take O the middle point of PQ and join CO, PQ will be parallel to the conjugate diameter to CO, but C is the middle point of PP' and therefore CO is parallel to $P'Q$, hence PQ, $P'Q$ are parallel to conjugate diameters.

15. The ratio of the rectangles of two intersecting chords of an hyperbola is the same when one or each is moved into any position parallel to its former position.

Let VP, Vp be the generating lines of the cone through the extremities of one of the chords POp, O being the point of intersection of the two chords. Through V the vertex draw VC parallel to POp of some fixed length, the same for all the chords. Let planes through O and C at right angles to the axis of the cone intersect the plane $PVCp$ (the plane of the paper) in the lines ROr, DdC; and the generating lines VP, Vp in R, r, D, d; then Rr, Dd are chords of the circles in which these planes cut the cone.

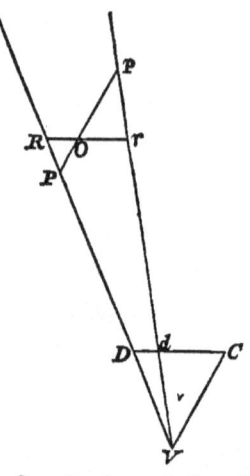

Also Ror is parallel to DdC as well as POp to VC.

Hence by similar triangles POR, VCD,
$$PO : RO :: VC : CD,$$
and by similar triangles pOr, VCd,
$$Op : Or :: VC : Cd;$$
therefore $PO . Op : RO . Or :: VC^2 : CD . Cd$.

Now let $P'Op'$ be another chord of the hyperbola with its extremities on different generating lines of the cone, intersecting the former chord in O. Draw VC' parallel to $P'Op'$ and of the same length as VC: also $R'Or'$, $D'd'C'$ the intersections of planes through O and C' at right angles to the axis with the plane $P'VC'p'$. Then $R'Or'$ is a chord of the same circle as ROr was, and $RO.Or=R'O.Or'$, also $VC = VC'$: and as before $P'O . Op' : R'O . Or' :: VC'^2 : C'D' . C'd'$.

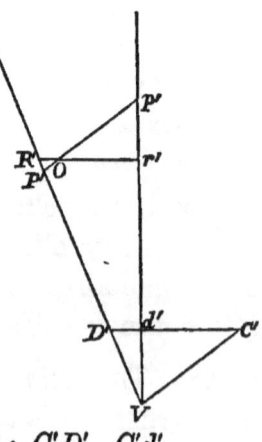

Hence we shall have

$$PO \cdot Op : P'O \cdot Op' :: C'D' \cdot C'd' : CD \cdot Cd.$$

Now observe that if the chords Pop, $P'op'$ are moved parallel to themselves into some new position, but so as to be still chords of the hyperbola, the lines VC, VC' will not be affected, but the generating lines on which the extremities of the chords rest will not be the same as before; yet C, C' remaining fixed, Dd, $D'd'$ in their new position will be still chords of the same circles as before, and each of the rectangles $CD \cdot Cd$, $C'D' \cdot C'd'$ will retain the same value as before. Hence the ratio $PO \cdot Op : P'O \cdot Op'$ will be invariable, and have the same value wherever O may be situated, provided the chords are drawn always parallel to their original position.

The same proof will hold when the extremities of the chords are on different branches of the curve, and when O is on the convex side of either branch.

The same proof will also hold for the other sections of the cone equally with the hyperbola.

16. The ratio of the rectangles on the segments of the chords equals that of the squares on the parallel semidiameters.

Let the chords move till they become tangents to the curve, viz. the tangents OP, OP' intersecting in O. The ratio of the rectangles is that of the squares on OP, OP'.

Draw CQ parallel to OP to meet the conjugate in Q, and draw QQ' parallel to PP' to meet CQ' parallel to OP' in Q'. Join CO, and let it meet QQ' in W, and when produced let it meet PP' in V.

Then V is the middle point of PP', and the triangles CWQ, CWQ' are similar to OVP, OVP':

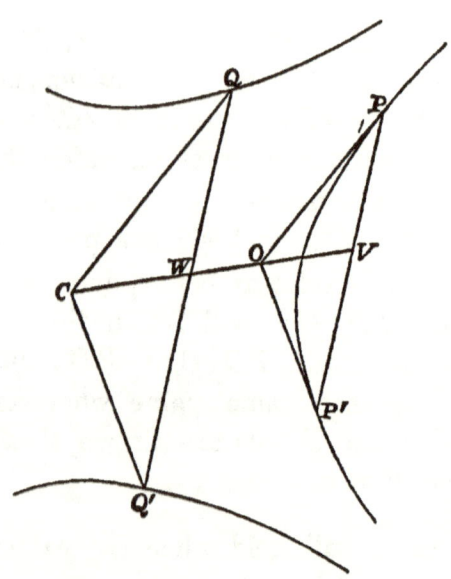

$$\therefore QW : WC :: PV : VO,$$
and $$Q'W : WC :: PV' : VO,$$
$$\therefore QW : Q'W :: PV : P'V,$$
but $PV = P'V$; therefore also $QW = Q'W$.

But CV bisecting PP', bisects all chords of the hyperbola and its conjugate parallel to PP', and therefore QQ', parallel to PP' and bisected by CO, is a chord of the conjugate hyperbola; CQ' is the semi-diameter parallel to OP', and the ratio of the rectangles under the segments of the chords $= OP^2 : OP'^2 = CQ^2 : CQ'^2$, since the triangles OPP', OQQ' are similar.

See also Besant's *Elementary Conic Sections*, p. 116.

COR. If a circle intersect an hyperbola in four points, it may be easily shewn that the diameters parallel to the pairs of opposite sides of the quadrilateral formed by drawing the common chords will be equal, and therefore equally inclined to the axes of the curve; hence the opposite sides of the quadrilateral will be equally inclined to the axes.

17. $\qquad NG : NC :: BC^2 : AC^2.$

Let PG be the normal at P; draw NR from the foot of the ordinate PN to touch the auxiliary circle at R; then if

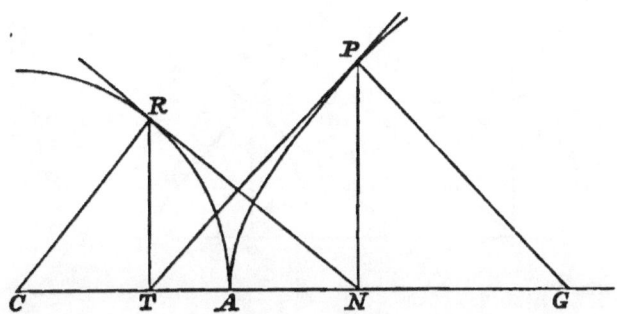

PT be the tangent at P, RT will be at right angles to the transverse axis, because $CN \cdot CT = CA^2$. Also

$$NR^2 = NA \cdot NA' \text{ and } \therefore NP^2 : NR^2 :: BC^2 : AC^2,$$

and $\qquad NP : NR :: BC : AC.$

Join CR: then in the right-angled triangles TPG, CRN, $PN^2 = TN \cdot NG$, $RN^2 = CN \cdot TN$, $\therefore PN^2 : RN^2 :: NG : NC$;

$$\therefore NG : NC :: NP^2 : NR^2 :: BC^2 : AC^2.$$

18. $\qquad PF \cdot PG = BC^2, \quad PF \cdot Pg = AC^2.$

If the normal at P intersects the axes in G and g, and PF be the perpendicular on CD the semi-diameter conjugate to CP, then first, $\qquad PF \cdot Pg = AC^2.$

J. C. S.

Let PT be the tangent at P, and let Pnr parallel to transverse axis cut CD and the conjugate axis in n and r,

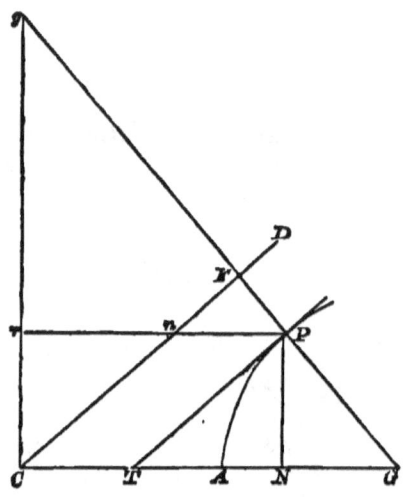

nT, rN will be parallelograms. Then since the angles at F and r are right angles, a circle will pass through F, n, r and g.

Hence $\quad PF \cdot Pg = Pn \cdot Pr = CT \cdot CN = CA^2$.

Also $\quad PG : Pg :: NG : NC :: BC^2 : AC^2$:

and $\quad PF \cdot Pg = AC^2$; $\therefore PF \cdot PG = BC^2$.

19. The asymptotes of any hyperbola are parallel to the generating lines of the cone in which a plane through the vertex parallel to the cutting plane cuts the cone.

Let VE be the line in which a plane through V parallel to the cutting plane cuts the plane of the paper.

Make VE equal to AC and draw FEF' through E at right angles to the axis.

EXAMPLES.

Then the part of the perpendicular through E to the plane of the paper between that plane and the cone will be the height of a right-angled triangle which has half the angle between two generating lines at the base, call it EG. Then $EG^2 = EF' \cdot EF''$: and from equal triangles ACD, VEF and

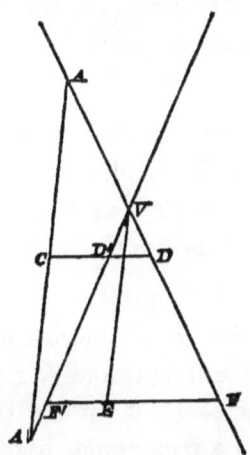

$A'CD'$, VEF'', $EF = CD$, $EF'' = CD'$; $\therefore EG^2 = CD \cdot CD' = BC^2$ and $EG = BC$. Hence the generating lines are inclined to VE at the same angle as the asymptotes to CA parallel to it, and being in a plane parallel to that containing the asymptotes, are parallel to them.

Examples.

1. Of all hyperbolas that can be cut from a given cone, the ratio $BC : AC$ is greatest in that cut by a plane parallel to the axis.

2. Give a construction for determining, if possible, the direction of the plane that will cut a rectangular hyperbola from a given cone.

EXAMPLES.

3. NPQ is drawn from any point N of the conjugate axis of a rectangular hyperbola at right angles to it, to cut the auxiliary circle and the hyperbola in P and Q; prove $NP^2 + NQ^2 = 2AC^2$.

4. Prove the same for any hyperbola and the ellipse on the same axes.

5. A circle passes through A, A', the vertices of a rectangular hyperbola: the common chord parallel to AA' is a diameter of the circle.

6. The tangent PT to a circle intersects a fixed diameter in T, and TQ is drawn at right angles to the diameter of a length bearing a constant ratio to TP: as P moves on the circle, Q will move on an hyperbola which has the given circle for its auxiliary circle.

7. Through any point P of a circle on AB as diameter PA, BP are drawn and produced to Q and R such that QR is at right angles to BA and bisected by it: prove that as P moves on the curve Q, R will move on a rectangular hyperbola.

8. Draw a tangent to an hyperbola parallel to a given line.

9. From a given point in an hyperbola draw a line such that the intercept between the other point of intersection and an asymptote shall equal a given line. When does the problem become impossible?

10. The tangent at P meets an asymptote in T, and TQ is drawn to the curve parallel to the other asymptote. PQ produced both ways meets the asymptotes in R, R'. RR' is trisected in P, Q.

11. From a given point P in a rectangular hyperbola PM, PN are drawn equally inclined to an asymptote, and when produced meet the curve again in Q, R; prove that QR is a diameter.

12. Find the position and magnitude of the axes of an hyperbola which has a given line for asymptote, touches another line in a given point, and passes through another given point.

EXAMPLES. 69

13. Draw lines from the centre of an hyperbola to the extremities of any chord: the intercepts of any line parallel to the chord between these lines and the asymptotes will be equal.

14. From $CV \cdot CT = CP^2$ deduce in the rectangular hyperbola, $VC \cdot VT = QV^2$.

15. In the rectangular hyperbola the triangles CVQ, QVT are similar.

16. In the rectangular hyperbola, R is the middle point of a chord Qq and RQ', Rq' are drawn parallel to the tangents at q, Q to meet CQ and Cq: shew that a circle will circumscribe $CQ'Rq'$.

17. From any point R of an asymptote RN, RM are drawn at right angles to the axes intersecting the hyperbola and its conjugate in P and D. Prove CP, CD are conjugate in the rectangular and general hyperbolas.

18. The tangent at P meets an asymptote in T, TN is drawn at right angles to the transverse axis; prove in the rectangular and general hyperbolas that NP passes through D the extremity of the diameter conjugate to CP.

19. Any two tangents have their points of intersection with the asymptotes joined: the lines so drawn will be parallel.

20. From any point P of an hyperbola PH, PK are drawn each parallel to one asymptote to meet the other: these lines produced if necessary meet any line through the centre in R and T. Complete the parallelogram $PRQT$, and shew that Q is a point on the curve.

21. Draw a tangent to an hyperbola at a given distance (less than AC) from C.

22. The tangent to an hyperbola meets a pair of conjugate diameters in T, t and the second tangents are drawn to the curve from T and t, they will touch it at the extremity of a diameter.

23. An hyperbola can be drawn through the extremities of any two radii of a circle having the diameters at right angles to the radii as asymptotes.

24. An hyperbola can be drawn through the extremities of any two semi-diameters of an ellipse having the diameters conjugate to them as asymptotes.

25. If PG be the normal at P, $CG = 2CN$ in the rectangular hyperbola.

26. $PG \cdot Pg = CD^2$.

27. If the tangent at P intersects the asymptotes in R, r the circle on Gg as diameter will pass through C, R and r.

28. In the same hyperbola Gg varies inversely as the perpendicular from the centre on the tangent.

29. An hyperbola is cut from a given cone, and a straight line drawn from a point of it parallel to an asymptote; the plane through the vertex of the cone and this line will cut the cone in two straight lines one of which is parallel to the line in the curve.

Hence prove the proposition of Art. 11.

30. Apply the method of proof in Art. 15 to shew that if parallel tangents at Q, Q' meet the tangent at P in T, T',

$$QT : PT :: Q'T' : P'T'.$$

Prove the following propositions in the rectangular hyperbola:

31. The lines bisecting the angles between CP and the tangent at P are parallel to the asymptotes.

32. The tangent at the point Q intersects a pair of conjugate diameters in T, T': prove that CQ is the tangent to the circle round CTT'.

EXAMPLES.

33. If V be the middle point of a chord, the lines bisecting the angles between CV and the chord are parallel to the asymptotes.

34. The lines bisecting the angles between supplemental chords are parallel to the asymptotes.

35. The angle between lines drawn from two points on the curve to one extremity of a diameter equals or is the supplement of that between the lines from the same points to the other extremity.

36. Diameters at right angles are equal to one another.

37. Of two chords of the curve at right angles to one another one has its extremities on the same branch, the other on different branches.

38. If AB, CD are chords at right angles to one another and the circle ABC cuts CD produced in E, BA will bisect ED.

39. The four circles that may be drawn through three of the points A, B, C, D in the last example consist of two pairs of equal circles.

40. If a tangent be at right angles to a chord, the circle on the chord as diameter will pass through the point of contact.

41. The line joining one end of a diameter with one end of a chord at right angles to it is at right angles to the line joining the other end of the diameter with the other end of the chord.

CHAPTER VII.

On the Focal properties of the Hyperbola.

1. LIKE the Ellipse, the Hyperbola has two foci and directrices: before treating of them let us premise the following propositions:

If the circles escribed on the sides VA, VA' of the triangle VAA' touch them and the produced sides in $L, S, K,$

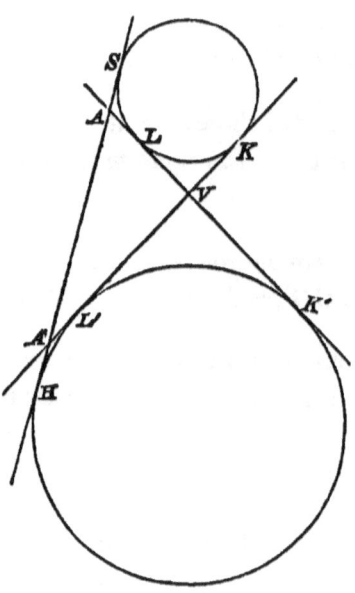

L', H, K' respectively, then $AS = A'H$ and KL' or $K'L$ $= AA'$.

For $SH = SA + AH = AL + AK'$
$\quad\quad = 2AL + LK'$,
also $= SA' + A'H = A'K + A'L' = 2A'L' + KL'$:
and $LK' = KL'$, $\quad \therefore AL = A'L'$ or $AS = A'H$.

Add AA' to each of these equals,

$\therefore AS + AA' = AH = AK' = AL + LK' = AS + LK'$;

$\therefore AA' = LK'$ or KL'.

Also, if we produce KL, $K'L'$ to meet AA' in X, X' (as in the figure of the next article), $AX = A'X'$.

For since $AL = A'L'$ and these lines are equally inclined to the line joining the centre of the circles, their projections on this line will be equal: and AX, $A'X'$ have the same projections on the same line as AL, $A'L'$ and therefore have their projections on it equal, and being in the same straight line are themselves equal.

2. Now let AA' be the line in which a plane perpendicular to the plane of the paper cuts it. Let AP, $A'P$ be the two branches of the hyperbola in which the same plane cuts the two sheets of the cone AVK, $A'VK'$ whose axis is the line joining the centres of the circles in the last Article: and let the circles SLK, $HL'K$ revolve about the axis, they will form spheres that touch the cone in circles LRK, $L'R'K'$.

Let P be any point in the hyperbola: the generating line PV of the cone will touch the spheres, in points R, R' suppose. Join SP, HP; these lines lie in the plane that touches the two spheres at S and H, and therefore touch them in those points: hence $SP = PR$, $HP = PR'$;

∴ $HP - SP = PR' - PR = RR' = KL' = AA'$, a distance independent of the position of the point P: hence the differ-

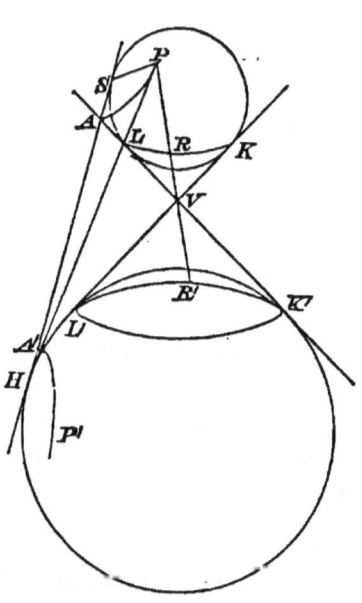

ence of the distances of any point in the hyperbola from S and H is constant. S and H are called the *foci*.

3. Now let the plane through P perpendicular to the axis of the cone cut the cone in the circle QPQ' and the plane of the hyperbola in NP at right angles to $A'A$ produced. Then we shall have $SP = PR = QK$, and the triangles QAN, KAX are similar;

∴ $QK : NX :: AK : AX :: AS : AX$.

Hence, wherever P is situated on the curve $SP : NX$ is a constant ratio $= AS : AX$.

Similarly, joining HP,
$$HP = PR' = QK',$$
and the triangles QAN, $K'AX'$ are similar;
$$\therefore QK' : NX' :: AK' : AX' :: AK : AX :: AS : AX.$$

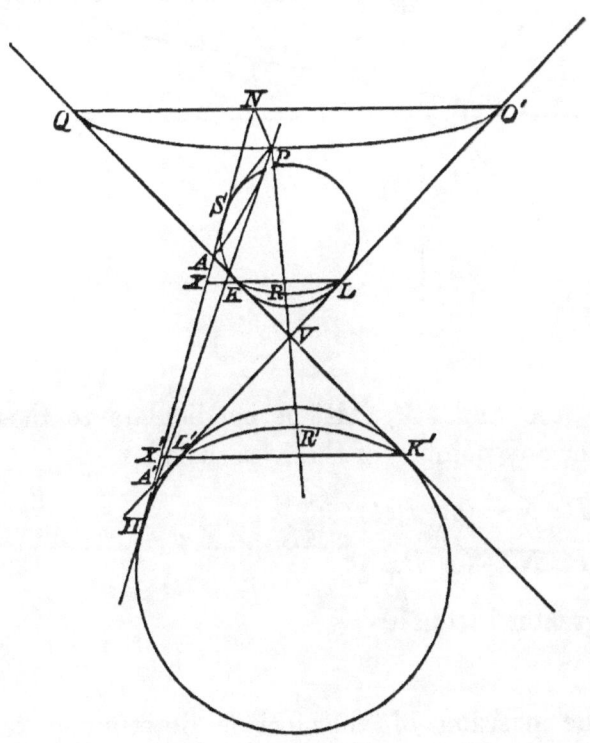

Hence also $HP : NX'$ is a constant ratio the same as $SP : NX$, wherever P is situated on the curve. We have shewn that $HA' : A'X' = SA : AX$: these ratios = $A'L' : A'X'$ or $AK : AX$; and the triangles AXK, $A'X'L'$ are both right angled at X or one of them has an obtuse angle at X; in either case we see that the ratios are ratios of greater inequality. As in the ellipse either of these ratios is called the *eccentricity*.

So then if we draw the hyperbolic section in the plane of the paper and draw through X, X' lines XZ, $X'Z'$ at right

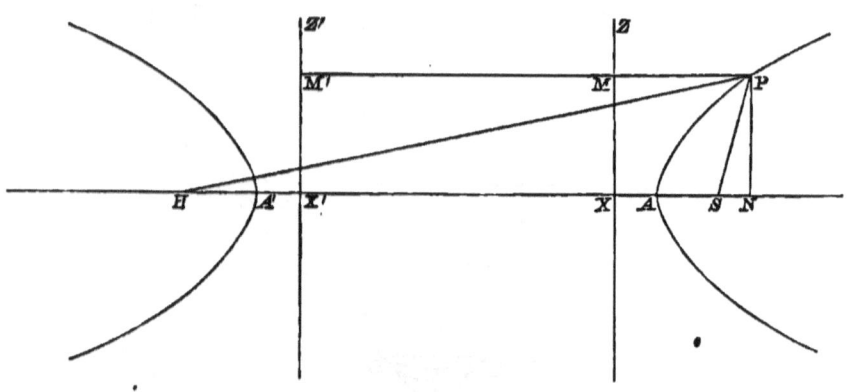

angles to XX' and PM, PM' perpendiculars to these lines, we have for any point P of the hyperbola

$$\frac{SP : NX \text{ (or } PM)}{HP : NX' \text{ (or } PM')} :: AS : AX :: A'H : A'X'$$

ratios of greater inequality.

4. The position of the foci is determined relatively to the magnitude of the axes by the following relation $CS^2 = CA^2 + CB^2$.

We have shewn that by giving the proper angle to the cone any hyperbola of given axes may be cut from it by a plane parallel to the axis. Let then the given hyperbola be made by the intersection of a plane parallel to the axis of the cone. The foci will still be the points where the inscribed spheres touch the cutting plane, and these spheres will be equal. The figure represents the section of the cone and spheres by the plane of the paper.

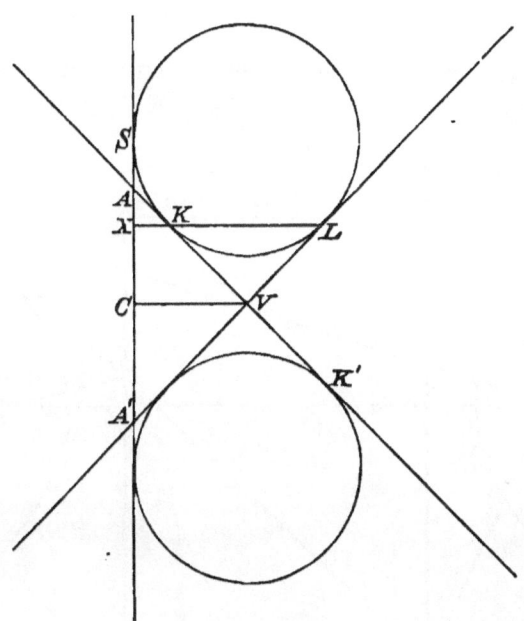

CV will now be perpendicular to the axis and equal to CB. Also $AA' = KK' = 2KV$ since the circles are equal:
$\therefore CA = KV$.

Hence $CS = CA + AS = KV + AK = AV$,
$\therefore CS^2 = AV^2 = CA^2 + CV^2 = CA^2 + CB^2$.

5. CX is a third proportional to CS and CA.

For in the last figure KX is parallel to CV,
$$\therefore AV : AC :: KV : XC,$$
$$\text{or } CS : CA :: CA : CX.$$

Cor. Each of these ratios $= AK : AX$ or $AS : AX$, the eccentricity.

6. Properties of the secant and tangent.

Art. 6. of Chap. V. applies verbatim to all the Conic Sections.

7. The focal distances make equal angles with the tangent at any point.

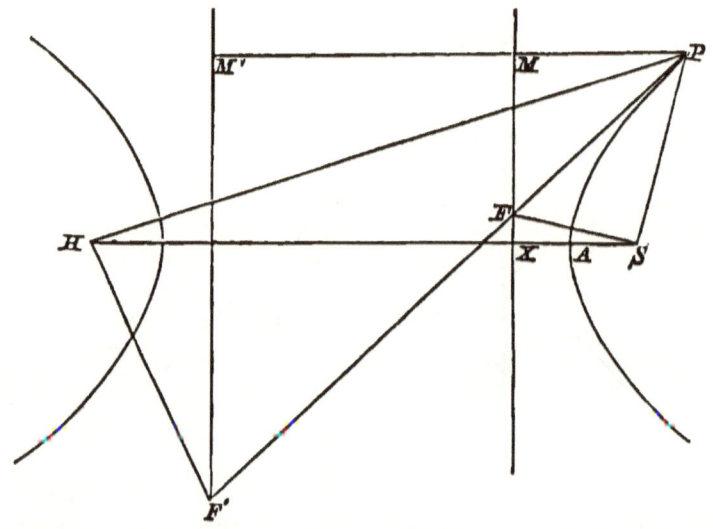

Let the tangent at P meet the directrices in F, F', join SF, HF'; PSF, PHF' will be right angles.

Also the triangles MPF, $M'PF'$ are similar:
and $SP : PM :: AS : AX :: HP : PM'$,
∴ $SP : PH :: PM : PM' :: PF : PF'$,
or $SP : PF :: HP : PF'$.

Hence (Euclid VI. A.) SPF, HPF' are similar triangles, and the angles SPF, HPF' are equal.

Cor. The tangents at A, A' the vertices of the curve are at right angles to the transverse axes.

ON THE FOCAL PROPERTIES OF THE HYPERBOLA. 79

8. The feet of the perpendiculars from the foci on any tangent lie on the circle on AA' as diameter.

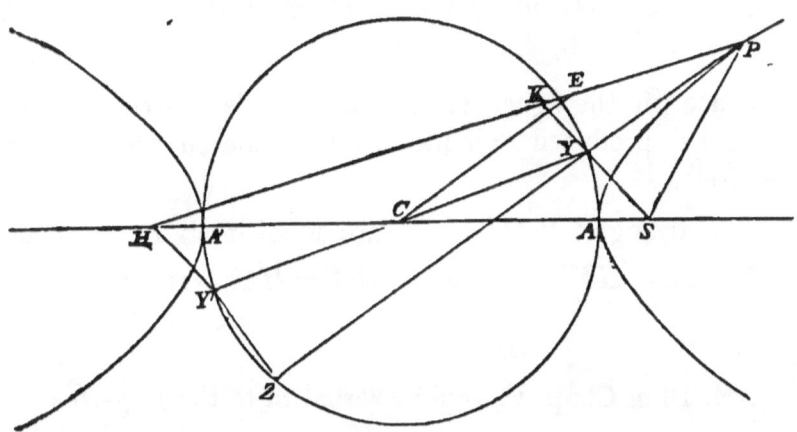

Let SY, the perpendicular from S on the tangent at P, meet HP in K: join SP, CY. Then in the right-angled triangles SYP, KYP, PY is common and the angle SPY = the angle HPY (by the last proposition).

Hence $SY = KY$, and $SP = KP$;

$$\therefore KH = HP - KP = HP - SP = 2AC.$$

And C, Y being the middle points of SH, SK, CY is parallel to HK and half of it and therefore $= AC$.

Hence Y lies on the circle on AA' as diameter. And similarly Z the foot of the perpendicular from H.

COR. 1. CY being parallel to HP and bisecting SH, when produced will also bisect SP. Hence SYP being a right angle, the circle on SP as diameter passes through Y and has its centre on CY produced; hence it touches the auxiliary circle at Y.

COR. 2. If CE be drawn parallel to the tangent at P

and therefore conjugate to CP and intersect PH in E, then $CEPY$ is a parallelogram and $PE = CY = AC$.

9. If SY, HZ are perpendiculars on the tangent at P from the foci $S, H,$ $SY.HZ = BC^2$.

Since (in the figure of the last Article) YZH is a right angle, YC produced will meet ZH on the circumference of the circle.

The triangles SCY, HCY' are equal in all respects, and $SY.HZ = HY'.HZ = HA.HA' = HC^2 - A'C^2 = BC^2$.

10. $\qquad\qquad SP.HP = CD^2$.

Art. 10 of Chap. V. applies verbatim to the hyperbola.

11. If from any point Q of the tangent PK perpendiculars QN, QT be drawn to the directrix and SP, then
$$ST : QN :: AS : AX.$$

12. Hence if QP, QP' be the tangents to an hyperbola from an exterior point Q, QP, QP' subtend equal angles at S.

See Articles 11, 12 of Chap. V.

If P, P' are not on the same branch of the curve, since Q lies on the near side of the directrix to one branch and the further side to the other, T or T' will lie in PS or $P'S$ produced, and the supplement of one of the angles QSP, QSP' will be equal to the other, i.e. the angles will be supplementary.

Cor. Combining this last proposition with that of Art. 7 we see that the point Q is equidistant from the four lines SP, HP, SP', HP'.

13. If QP, QP' be the tangents to an hyperbola from a

point Q, the angles SQP, HQP' are supplementary, as are also SQP', HQP.

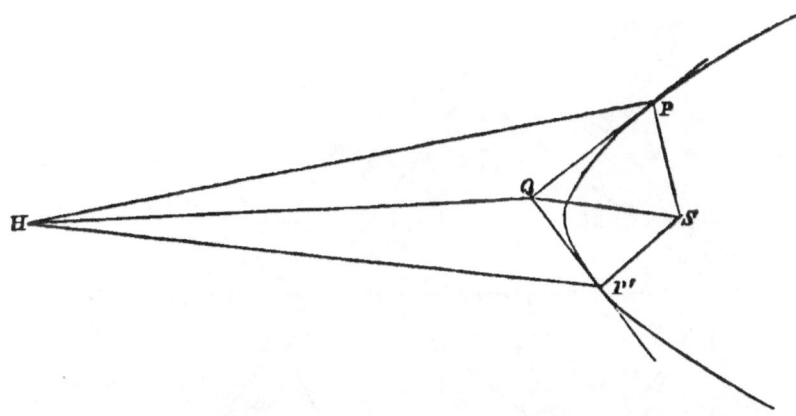

Join SP, SQ, SP', HP, HQ, HP'.

The angles of the quadrilateral $SPHP'$ are bisected by the lines drawn from the angular points to Q;

∴ the sum of the angles SPQ, PSQ, $HP'Q$, $P'HQ$ = sum of $P'SQ$, $SP'Q$, PHQ, HPQ.

But the first sum of angles with SQP, HQP' = the sum of the interior angles of two triangles = the second sum with SQP', HQP: hence SQP, HQP' together = SQP', HQP together: and these four angles together = four right angles: hence $SQP + HQP' = SQP' + HQP$ = two right angles. Q.E.D.

If P, P' lie on different branches it may be shewn, as in the case of the Ellipse, that SQP, HQP' are each half of the angle between HP and SP', and therefore (no longer supplementary but) equal to one another.

14. Any two tangents at right angles to one another intersect on a circle whose centre is C and square on the radius $+ BC^2 = AC^2$.

Let any two tangents at right angles to one another intersect in Q and cut the circle on AA' as diameter in Y, Z;

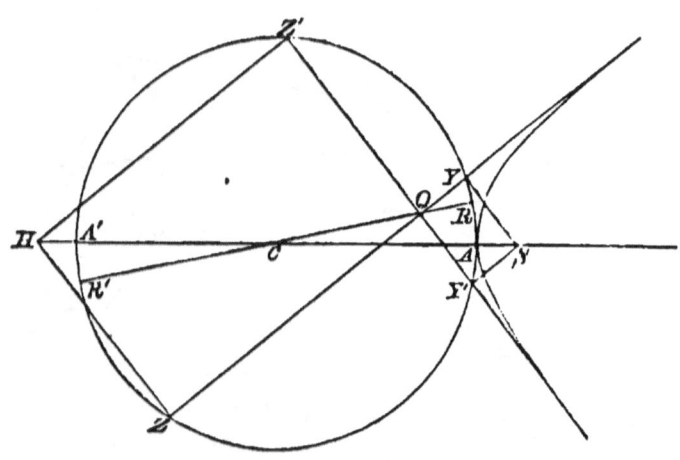

Y', Z'. Draw SY, SY'; HZ, HZ'. Let QC cut the circle in R, R'.

Then SQ, HQ will be rectangles with opposite sides equal.

Hence $QR \cdot QR' = QY' \cdot QZ' = SY \cdot HZ = BC^2$,

and $CQ^2 + QR \cdot QR' = CR^2$;

$\therefore CQ^2 + BC^2 = AC^2$:

and the locus of Q is a circle with centre C and square on its radius = difference of the squares on AC and BC.

Cor. If AC be less than BC the curve has no tangents at right angles to one another: but in this case the circle will be the locus of the intersections of tangents to the conjugate hyperbola at right angles to one another.

15. $SG : SP :: SA : AX.$

The tangent at P bisects the angle SPH: hence the

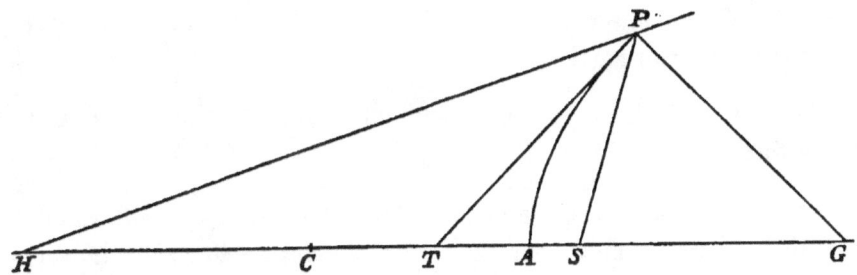

normal PG bisects the angle between SP and HP produced. Hence (Euclid VI. A)

$$SG : SP :: HG : HP :: HG - SG : HP - SP$$
$$:: 2SC : 2AC :: CS : CA :: SA : AX.$$

EXAMPLES.

1. Given a focus and two points of an ellipse, the other focus lies always on a certain hyperbola.

2. Four circles are drawn having the corners of a square for centres and diameters equal to the side: the locus of the centres of all circles which touch opposite pairs of circles, one internally and the other externally, is a pair of rectangular hyperbolas which have the common tangents of the circles for asymptotes.

3. The tangent from C to the circle through the foot of the directrix and the extremities of the latus rectum equals CS.

4. If N be the middle point of SX, prove $AN . A'N = SN^2$.

5. The portion of either asymptote between the directrices equals the transverse axis.

6. Given the asymptotes and directrices, find the foci.

7. Given one asymptote and the direction of the second and a focus, find the vertices.

8. Given the asymptotes and one point on the curve, find the foci.

9. The focal distance of any point P on the hyperbola equals a line drawn from P parallel to an asymptote to meet the corresponding directrix.

10. Each of the tangents drawn to the auxiliary circle from the foci equals BC and touches it in a point where the directrix cuts it.

11. Two hyperbolas have the same asymptotes, shew that the chord of one touching the other is bisected at the point of contact.

12. Two hyperbolas have their foci coincident and the angles between the asymptotes supplementary: no tangent to one can be at right angles to a tangent to the other except the asymptotes.

13. The line drawn from S parallel to an asymptote to meet the curve equals a quarter of the latus rectum.

14. Two chords through the same focus have three of their four extremities on one branch, and the remaining extremity on the other: prove that the four lines joining their extremities intersect two and two on the corresponding directrix.

15. PQ a chord of one branch of an hyperbola subtends at S an angle double of that subtended at S by a length pq on the corresponding directrix: shew that Pp, Qq intersect on the curve.

16. An ellipse cuts any confocal hyperbola at right angles.

EXAMPLES.

17. A circle is drawn through the foci and any point P of an hyperbola, the tangent and normal at P meet the conjugate axis in the same points as this circle.

18. P, Q are points in two confocal hyperbolas at which SH subtends equal angles, the tangents at P, Q are inclined at an angle equal to that subtended by PQ at either focus.

19. HY' drawn parallel to SP meets SY produced on the circumference of a fixed circle.

20. Given one focus, one point and one tangent of an hyperbola, the locus of the other focus is an hyperbola.

21. Qq a chord of the asymptotes moves parallel to itself and tangents are drawn from Q, q, their intersection will lie on a straight line through C.

22. A rectangular hyperbola confocal with an ellipse cuts it at the extremities of equal conjugate diameters of the ellipse.

23. An ellipse and confocal hyperbola intersect in P: one asymptote passes through the point of the auxiliary circle of the ellipse corresponding to P.

24. An ellipse and hyperbola have each their foci at the vertices of the other: if the tangents at the point of intersection meet the conjugate axis in t, t', $Ct = Ct'$.

25. The tangent at P is perpendicular to one asymptote and PS through the focus meets that asymptote in Q, prove $SQ = AC$.

26. P is the point of the hyperbola where SP is at right angles to HP: and they intersect CD in E, E'.

Prove $EE'^2 = 2AC^2$ and $CD^2 = 2BC^2$.

27. From the point of intersection of an asymptote and directrix a tangent is drawn to the hyperbola: the line joining the interior focus of the branch it touches with the point of contact is parallel to the asymptote.

28. The tangent from P a point in the asymptote touches the curve in O: HT is parallel to the same asymptote; prove that HP bisects the angle THO: and if PQ intersects the other asymptote in Q, $PHQ=$ half the angle PCQ.

29. The focal distances of two points P, P' intersect in O, prove that the tangents QP, QP' subtend equal angles at O.

30. If SP, HQ are parallel, find the locus of the intersection of the tangents at P and Q.

31. PT, QT tangents from T a point in the auxiliary circle whether to the same branch or not have one of the focal distances SP, SQ parallel to one of the two HQ, HP.

32. The tangent at P is perpendicular to two parallel tangents at Q, Q', prove that SQ, HQ' subtend equal angles at P.

33. If GL be the perpendicular from G on SP, prove $GL:PN$ a constant ratio.

34. From the vertex A draw AQ perpendicular to the tangent at P, and let QA produced meet PS produced in O, the locus of O is a circle.

35. Let GY produced meet ZH produced in Z', then $HZ'=HZ$.

36. If the normals at P, Q extremities of a focal chord intersect in O, and OL parallel to the transverse axis cut SP in L, L is the middle point of PQ.

37. If GR be the perpendicular on SP from G, PR equals semi-latus rectum.

38. If the normal at P meets the axes in G, g, the triangles SPg, GSg are similar, and $Gg:Sg$ and therefore $Pg:Sg$ are constant ratios.

CHAPTER VIII.

On the Parabola.

1. WE have now to consider the case in which the cutting plane is parallel to one of the generating lines of the cone. Let the cutting plane intersect the cone in

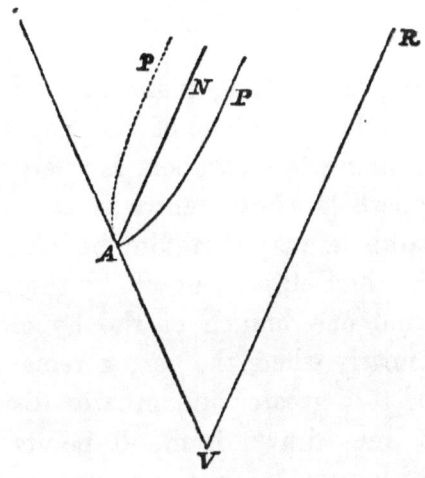

the curve PAp, and intersect the plane of the paper which contains the axis in the line AN parallel to the generating line VR in the plane of the paper, then PAp is called a *Parabola*.

It is manifest that the parabola is divided into equal parts by the line AN which is called the axis, every line in the plane of the section at right angles to AN will meet the curve in two points on opposite sides of AN and at equal distances from it.

The axis intersects the curve in one point only, in other words the curve has only one vertex.

2. By turning the cutting plane about a line through A at right angles to the plane of the paper through the smallest possible angle, the parabola is changed either into an ellipse or hyperbola whose centre is at a great distance from C: the parabola may therefore be considered as the form to which the semi-ellipse cut off by the axis minor on the side of A and one branch of the hyperbola approach more and more nearly when, the vertex remaining fixed, the centre is removed to a greater and greater distance. Hence it appears that lines drawn from all points in the semi-ellipse and semi-hyperbola to their centres become more and more nearly parallel as the cutting plane moves towards the position in which it cuts a parabola from the cone.

And we may anticipate that the line joining the middle points of parallel chords of a parabola will be parallel to its axis; and generally all properties of the ellipse and hyperbola that relate to lines that remain finite in the parabola are equally true in that curve.

It will have been observed that there is a similarity in the properties of the ellipse and hyperbola; the parabola is the curve in which they meet, or in which each undergoes the transition into the other.

3. $NP^2 = 4AS \cdot AN$.

Let RPR' be a circular section of the cone made by a plane at right angles to the axis through N, any point of

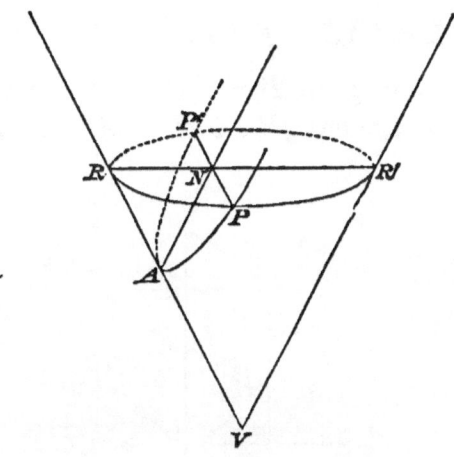

the axis of the parabola and cutting the plane of the curve in PNP' at right angles to the axis.

Then $NP^2 = RN \cdot NR'$, and the ratio $NR : AN$ is constant for all positions of N.

Hence the ratio $NP^2 : AN \cdot NR'$ is constant: and NR' is also invariable.

The ratio $RN : AN$ is invariable for all positions of N; the ratio $NR' : AN$ may be made to have all values from a ratio indefinitely great to one indefinitely small by moving N along the axis from A. Hence we may find a point S in the axis such that for that point $RN \cdot NR' = 4AN^2$, or if LSL' be the double ordinate through S, $SL^2 = 4AS^2$ or $SL = 2AS$.

Hence generally

$NP^2 : SL^2 :: AN . NR' : AS . NR' :: AN : AS$,
$NP^2 : 4AS^2 :: AN : AS$,

$$\therefore NP^2 = 4AS . AN.$$

4. $QR . Q'R = 4AS . PR$.

Let PR from any point P of the curve meet the double ordinate QMQ' in R, then $QR . Q'R = 4AS . PR$. Draw the

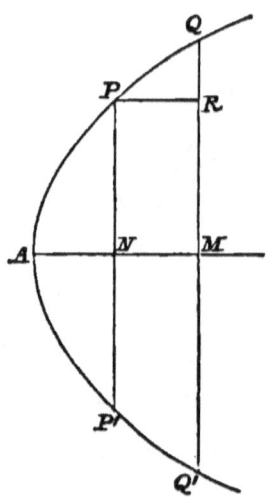

double ordinate PNP', then PM is a rectangle whose opposite sides are equal: and since QQ' is bisected in M,

$$QR . Q'R + RM^2 = QM^2,$$
or $QR . Q'R + PN^2 = QM^2$;
$\therefore QR . Q'R + 4AS . AN = 4ASAM$;
$\therefore QR . Q'R = 4AS . MN$
$= 4AS . PR.$

This proposition is equally true when, as in fig. § 6, R is a point in the chord QQ' external to the curve.

This property is seen in the case of the hyperbola in Chap. V. § 11. By Chap. VI. § 19 it appears that as the plane that cuts an hyperbola from the cone moves about the line through A towards the position in which it cuts a parabola, the angle between the asymptotes diminishes and they become more and more nearly parallel to the axis. Hence, as the hyperbola passes into the parabola, the line from any point of the curve parallel to an asymptote becomes parallel to the axis: and the proposition of the present Article takes the place of that previously proved for the hyperbola.

5. The middle points of parallel chords lie on a straight line parallel to the axis.

Let V be the middle point of any chord PQ, draw the

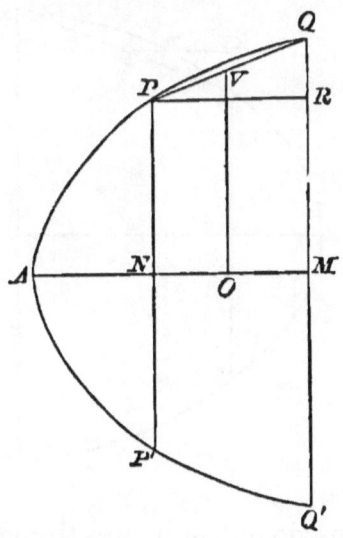

ordinates PN, VO, and the double ordinate QMQ', PR parallel to the axis.

Then since QQ' is bisected in M,
$$RQ' = RM + MQ' = PN + QM = 2VO,$$
and $QR \cdot RQ' = 4AS \cdot PR$ or $QR \cdot VO = 2AS \cdot PR$,
and $VO : 2AS :: PR : QR$ a constant ratio for all parallel chords. Hence VO is invariable for parallel chords, and all their middle points lie on a straight line parallel to the axis.

This line is called the *diameter* of the chords.

This property corresponds to the fact that the ellipse and hyperbola have the middle points of all parallel chords on a straight line through the centre.

6. $RQ^2 = 4AS \cdot PV$.

From V the middle point of any chord Qq draw VP parallel to the axis to meet the curve in P. Let PV produced

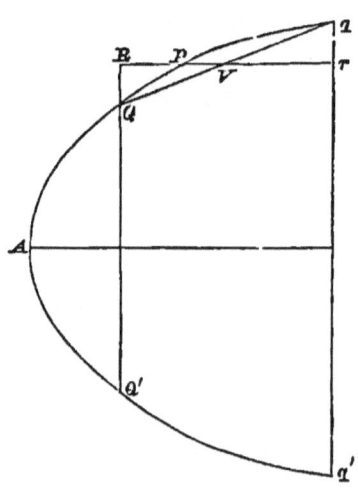

both ways meet the double ordinates through Q, q (the former produced) in R, r: V will be the middle point of Rr,
and $Pr = PV + Vr = PV + VR = PR + 2PV$.

Also $qr = QR$, and $q'r = Q'R + 2QR$;

$$\therefore qr \cdot q'r = QR \cdot Q'R + 2QR^2,$$

or $\quad 4AS \cdot Pr = 4AS \cdot PR + 2QR^2,$

$$QR^2 + 2AS \cdot PR = 2AS \cdot Pr$$
$$= 2AS \cdot PR + 4AS \cdot PV;$$
$$\therefore QR^2 = 4AS \cdot PV.$$

7. The point S we shall see is the focus of the curve, and we may now conveniently consider the properties of the curve with relation to it.

We premise the following proposition. If a circle touch the parallel lines VL, AS, and the line AV that intersects them in L, S and K, then if LK produced meets SA produced in X, $AS = AX$.

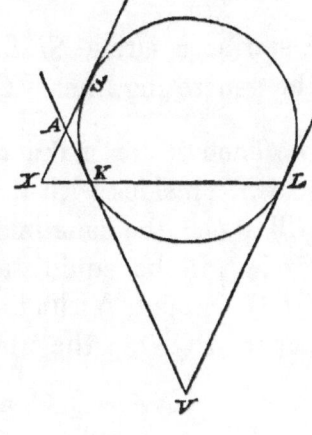

For $VK = VL$,

$\therefore$ the angle $VLK = VKL$: but $VLK =$ alternate angle AXK, and $VKL =$ vertical opposite angle AKX:

$\therefore AXK = AKX$ and $AK = AX$, but $AS = AK$;

$\therefore AS = AX$.

8. Now let AN be the intersection of the plane of the paper with a plane at right angles to it that cuts the cone

QVQ' in the parabola PAP', so that AN is parallel to VQ'.

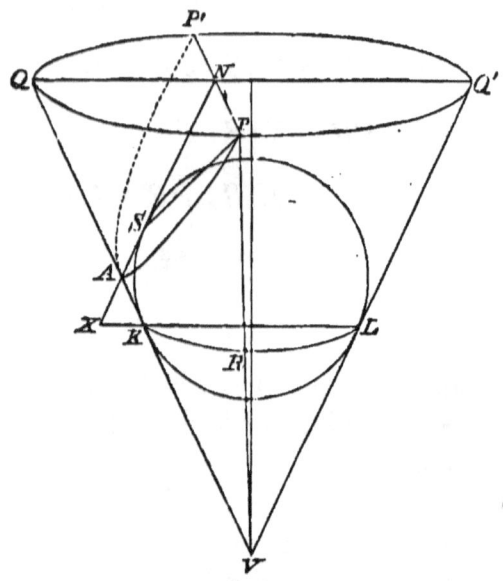

Describe a circle SKL in the plane of the paper touching the generating lines VQ, VQ' in K and L, and AN in S.

Then if we make the circle revolve about its diameter which coincides with the axis, it will generate a sphere which will touch the cone in the circle KRL. Every point of this circle will be equidistant from V. Let a circular section QPQ' through N cut the parabola in P and the plane of the paper in QNQ': the triangle QAN is similar to QVQ':

$$\therefore NA = QA: \text{ and } AX = AK, \therefore NX = QK.$$

Join SP; also PV cutting the circle KRL in R. PR will $= QK$. And SP, PR tangents drawn to the sphere SKL from the point P are equal : $\therefore SP = PR = QK = NX$: hence the distance of any point of the curve from $S =$ the distance of the foot of its ordinate from X.

So then if we draw the parabolic section in the plane of the paper and draw through X a line XM at right angles to the axis, and PM at right angles to XM, then we have for any point of the curve $SP = NX = PM$: the distance of any point P from $S =$ its perpendicular distance from XM. As before, S is the *focus* and XM the *directrix* of the parabola.

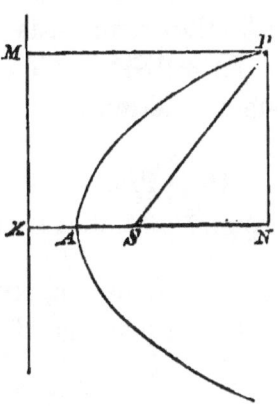

The eccentricity of the parabola is thus seen to be a ratio of equality. This results also from the fact that the parabola is the limiting form of a semi-ellipse, or of one branch of the hyperbola, when the centre moves to a continually increasing distance from the vertex, whilst the focus approaches a position at a certain definite distance from the vertex.

In this case $AS : AX$ tends to become a ratio of equality. For in the ellipse and hyperbola $CS \cdot CX = CA^2$; therefore if CS, CX are the distances of an external point C from the points where a line from C through the centre of a circle cuts the circle, CA will be the length of the tangent: and as C moves to a continually increasing distance, the tangent will become more and more nearly parallel to the diameter through C, and CA will tend to become an arithmetic mean between CS, CX. Hence as the semi-ellipse or branch of the hyperbola passes into the parabola, the vertex will assume the position midway between S and X, and the eccentricity will become a ratio of equality.

9. The position of S is identical with that previously assigned to it in proving the relation $PN^2 = 4AS \cdot AN$. For

SL the semi-latus rectum or ordinate through S will $= SX = 2SA$, which is the relation by which S was previously determined.

10. Properties of the secant and tangent. Art. 6 of Chap. IV. applies verbatim to the Parabola.

11. *The tangent at any point makes equal angles with the focal distance of the point and the axis of the curve.*

Let the tangent at P meet the directrix in F and the axis in T: join SP, SF, and draw the perpendicular PM. By the last Art. PSF is a right angle, also $SP = PM$; hence the right-angled triangles PSF, PMF having equal heights and hypotenuse common are equal in all respects: the angle $SPF = MPF = PTS$.

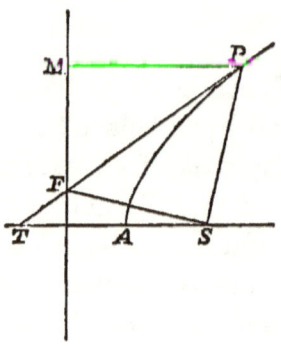

COR. 1. Hence $SP = ST$, and the perpendicular on the tangent from S will bisect the angle PST.

COR. 2. The angle between the focal distance and the tangent at any point cannot be a right angle except at the vertex, when SP, PM are in the same straight line and the tangent makes equal angles with them.

ON THE PARABOLA.   97

12. Tangents at the extremities of a focal chord intersect on the directrix in a right angle.

Let PF, QF, tangents at the extremities of the focal chord PSQ, intersect on the directrix in the point F and cut the axis in points T, T'.

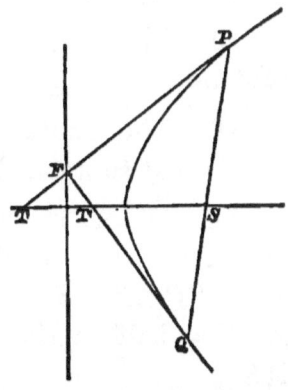

Then the angle PFQ
$$= FTT' + FT'T$$
$$= PTS + QT'S$$
$$= QPF + PQF$$
= half the sum of the interior angles of the triangle FPQ
= a right angle.

13. The foot of the perpendicular from the focus on any tangent, lies on the tangent at the vertex.

Let PR the tangent at P intersect SM in R. Then in the triangles SPR, MPR, $SP = PM$, and the angle $SPR = MPR$, hence the triangles are equal in all respects and the angle $SRP = MRP$, SM is at right angles to PR. Join AR; then SX is bisected in A and SM in R:

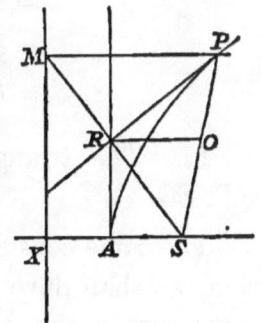

∴ AR is parallel to XM and is perpendicular to the axis and touches the parabola at A the vertex. Hence the foot of the perpendicular on PR from S lies on the tangent at A.

Cor. If O be the middle point of SP, RO will be parallel to MP, and therefore perpendicular to AR, and the circle on SP as diameter will touch AR at R. AR is the limiting form of the auxiliary circle in the ellipse and

J. C. S.

hyperbola when the centre removes to a continually increasing distance.

14. $ST : QN :: AS : AX$, as in Chap. IV. § 11.

15. The tangents to a parabola from an external point subtend equal angles at the focus, as in Chap. IV. § 12.

16. The angle between two tangents to a parabola equals that subtended by either tangent at the focus.

Let SY, SZ be the perpendiculars from the focus on the tangents to the parabola at P and Q which intersect in O. Join SP, SQ, SO. Then the angle $ASY = $ half ASP, and $ASZ = $ half ASQ.

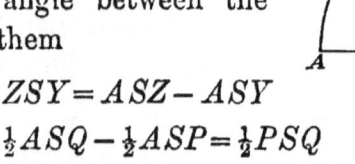

Also the angle between the tangents equals the angle between the perpendiculars on them

$$= ZSY = ASZ - ASY$$
$$= \tfrac{1}{2}ASQ - \tfrac{1}{2}ASP = \tfrac{1}{2}PSQ$$
$$= OSP \text{ or } OSQ.$$

17. The triangles SOP, SQO are similar and $SO^2 = SP . SQ$.

The same construction being made as in the last proposition, we shall have the sum of the angles SOP, $POZ = SOZ$ = the sum of the interior and opposite angles of the triangle SQO = the sum of SQO, OSQ.

But $POZ = OSQ$: ∴ $SOP = SQO$.

And the angle $PSO = OSQ$. Hence the triangles SOP, SQO are similar and $SP : SO :: SO : SQ$ or $SO^2 = SP . SQ$.

Cor. If SY be the perpendicular from S on the tangent at P, YA will be the tangent at A, and $SY^2 = SA . SP$.

ON THE PARABOLA. 99

18. The circle circumscribing the triangle formed by three tangents to a parabola passes through the focus.

Let two of the tangents intersect in O and the third tangent cut them in P, Q and touch the curve in R. Join SR and produce QO to Y.

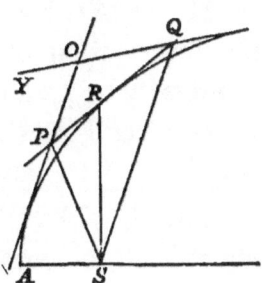

Then the angle $OPQ = PSR$,
and the angle $\quad OQP = QSR$:
hence the angle $\quad POY = OPQ + OQP$
$= PSR + QSR = QSP$:
the sum of the opposite angles POQ, PSQ of the quadrilateral $OPSQ$ = the sum of POQ, POY = two right angles, and the quadrilateral can be inscribed in a circle. In other words, the circle circumscribing the triangle POQ will pass through S.

19. The tangents at the extremities of a chord intersect on the diameter of the chord. See Chap. v. § 7.

20. $PT = PV$.

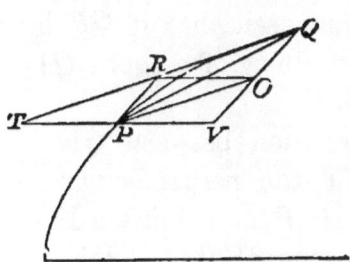

Let QT be the tangent at the point Q and QV the ordinate at that point to the diameter PV: to shew that $PT = PV$. For QV will be parallel to PR the tangent at P. Draw PO parallel to QT to meet QV in O. Join RO: being the diagonal of the parallelogram $POQR$ it will bisect the chord PQ, which is the other diagonal.

Hence RO bisecting the chord PQ and passing through the intersection of the tangents at the extremities of the chord will be parallel to the axis of the curve and therefore to PV: $RPVO$, $RTPO$ are parallelograms and $PT = RO = PV$.

COR. 1. If PN be the ordinate from any point P to the axis of the curve, and PT the tangent at that point, we shall have $AT = AN$.

COR. 2. To draw the tangents to the parabola from an external point T, draw TPV parallel to the axis meeting the curve at P; make PV equal to PT, and through V draw the chord QVq parallel to the tangent at P: QT, qT will be the tangents required.

The proposition of this Article results from the corresponding $CV \cdot CT = CP^2$ of the ellipse and hyperbola in the same way that $AS = AX$ results from the relation $CS \cdot CX = CA^2$.

21. $QV^2 = 4\ SP \cdot PV$.

We have already seen that if QR be the perpendicular distance of the extremity of a chord QVq from its diameter PV, $QR^2 = 4\ AS \cdot PV$.

To find the relation between QR and QV draw SY the perpendicular on the tangent at P and join AY which will be the tangent at A. Join SP. Then the angle $SPY = RPY =$ the interior and opposite angle RVQ, since QV is parallel to PY; the right-angled triangles SPY, QVR are similar and $QV^2 : QR^2 :: SP^2 : SY^2$. But $SY^2 = AS \cdot SP$;

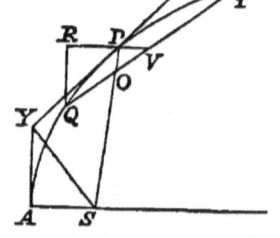

ON THE PARABOLA.

$$\therefore SP^2 : SY^2 :: SP^2 : AS \cdot SP :: SP : AS;$$
$$\therefore QV^2 : QR^2 :: SP : AS.$$

But $QR^2 = 4AS \cdot PV$, $\therefore QV^2 = 4SP \cdot PV$.

$4SP$ is called the *Parameter* of the diameter PV.

22. The length of the focal chord parallel to the tangent at $P = 4SP$.

Let SP intersect QV in O. Then QV being parallel to the tangent at P, makes equal angles with SP and PV; $\therefore$ the angle $POV = PVO$. $\therefore PV = PO$.

Hence if Qq passed through S we should have $PO = SP = PV$ and $QV^2 = 4SP^2$ or $QV = 2SP$, $Qq = 4SP$.

23. If two intersecting chords to a parabola move so as to be parallel to their former directions, the ratio of the rectangles under the segments is unaltered.

The method used in the case of the hyperbola (Chap. v. § 15) is equally applicable to the parabola. Or we may proceed thus:

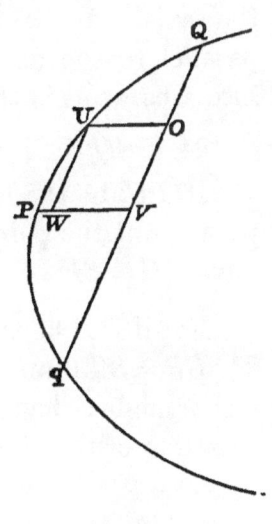

Let Qq be one of two chords which intersect in O, PV its diameter. Draw OU, UW parallel to PV, Qq.

Then $QV^2 = 4SP \cdot PV$, $UW^2 = 4SP \cdot PW$.

Or $QV^2 = 4SP \cdot PV$
$= 4SP \cdot PW + 4SP \cdot WV$
$= UW^2 + 4SP \cdot WV$
$= VO^2 + 4SP \cdot UO$.

But $QV^2 = QO \cdot qO + VO^2$, $\therefore QO \cdot qO = 4SP \cdot UO$.

If $Q'Oq'$ be another chord through O, $P'V'$ its diameter we shall have $Q'O \cdot q'O = 4SP' \cdot UO$.

Hence $QO \cdot qO : Q'O \cdot q'O :: SP : SP'$.

When Qq, $Q'q'$ move parallel to themselves into any other position, P and P', the extremities of their diameters, remain fixed and SP, SP' are constant. Hence the ratio $QO \cdot qO : Q'O \cdot q'O$ is constant.

COROLLARY. From the relation $QO \cdot qO = 4SP \cdot UO$ we see that if a system of parallel chords cut a line drawn from any point in the curve parallel to the axis, the rectangle of their segments made by that line will equal the rectangle of the intercepts on it and a constant line.

This property is analogous to that of the hyperbola in Chap. IV. § 11.

For by Chap. IV. § 19 we see that when the axis of the hyperbola becomes more and more nearly parallel to a generating line of the cone, the asymptotes become more and more nearly parallel to that line and therefore to the axis. Hence the limiting direction of a line drawn parallel to the asymptote of an hyperbola when it passes into a parabola is the direction parallel to the axis.

24. $SG = SP$.

If PG be the normal to the parabola at the point P we shall have $SG = SP$.

For if PT be the tangent at P, $SP = ST$; and TPG is a right angle: hence TG is a diameter of the circle with centre S and distance SP or ST: $\therefore SG = SP$.

This corresponds to the proportion that holds, in the

other sections of the cone $SP : SG :: AS : AX$: in the parabola $AS = AX$ and $SG = SP$.

25. $NG = 2AS$.

Let PN be the ordinate at P; PT, PG the tangent and normal at the same point. Then $NG = 2AS$.

For $\quad TN = 2TA, \ TG = 2TS.$
$\quad\quad\therefore \ TG - TN = 2TS - 2TA,$
or $\quad\quad NG = 2AS.$

Examples.

1. A series of parabolas pass through two given points, and the axis is always parallel to the same line: prove that the focus will lie always on a certain hyperbola.

2. If two confocal parabolas intersect, their common chord passes through the intersection of the directrices, and bisects the angle between them.

3. Two parabolas have a common directrix, prove that their common chord bisects the line joining their foci at right angles.

4. PSQ is a focal chord of a parabola: PA, QA meet directrix in Y, Z. Prove that PZ, QY are parallel to the axis.

5. The normals at the extremities of a focal chord intersect in K. KL is perpendicular to the chord, KF parallel to the axis. Prove that F is the middle point of SL.

6. The tangent at P is parallel to the focal chord QSQ'; PV is its diameter; prove that the normal at P bisects VS.

7. Given two tangents and the directrix, find the focus.

8. Tangents at P, Q extremities of a focal chord intersect on the directrix in T. Normals at P, Q cut TS in U, V. Prove that $PQ^2 = TU . TV$.

9. Given the focal chord PSQ and the focus S, find the vertex.

10. The normals at the extremities of the focal chord PSQ intersect in R. Shew that $PR^2 = SQ . QP$.

11. OP, OQ are tangents drawn to the parabola from O; OR parallel to the axis to meet the curve in R. The part of the tangent at R between OP, OQ is bisected at R.

12. OP, OQ are tangents drawn to the parabola from O, OT perpendicular to OP, cuts the normal at Q in T; join OS, ST: OST is a right angle.

13. PN is the ordinate of the point P of the parabola: SY the perpendicular from the focus on the tangent at P: prove $YP = YN$.

14. Find a point such that the tangents from it and the focal distances of the points of contact may be a parallelogram.

15. Two parabolas have the same focus and axes coincident, the line SPQ from the focus cuts them in P and Q: prove that the tangents at P and Q are parallel.

16. A circle through S touches the parabola at P. MPK drawn from the directrix parallel to the axis meets the circle again in K: prove that MSK is a right angle.

17. Two parabolas have a common directrix, prove that the common tangents intersect in it.

18. A given straight line is a chord of a parabola and at one end a normal: the axis is given in direction: find the focus and directrix.

19. If the tangent at any point P intersect the latus rectum SL in R, prove $SL : SR :: PN : SP$.

20. PQ is a tangent bounded by tangents OR, OT: PV, QV are drawn parallel to OT, OR: shew that V lies on RT.

21. The locus of the middle points of focal chords is a parabola whose latus rectum = half that of the original curve.

22. PV is the diameter of a focal chord QSQ', QD meets PV produced at right angles in D: prove that $VD = VS$, and $QS \cdot Q'S = QD^2$.

23. If the diameter OV of a focal chord meets the directrix in O prove that $SO^2 = 2AS \cdot OV$.

24. A circle touches the directrix at M and the diameter from M cuts the curve in P: the diameter of the circle $= 4SP$. Shew that the common chord passes through S and that MP produced bisects it.

25. Two parabolas with a common vertex are turned in opposite directions on the same axis, and the focus of one is eight times as distant from the vertex as that of the other: the common tangent is drawn at the vertex. Every tangent to the first parabola has the part between the tangent at the vertex and the axis bisected by the other.

26. If a circle touch a parabola at P and cut it in two points Q, R: the tangent at P and the chord QR are equally inclined to the axis: and PO drawn to the point where the axis cuts the circle is parallel to QR, and therefore $PQ = OR$.

27. In a parabola two chords are equally inclined to the axis; if another parabola passes through the extremities of these chords, it will have its axis at right angles to that of the first parabola.

28. Two parabolas have a common focus and axis; their vertices are turned in opposite directions: a straight line from S cuts them in P and Q: prove that the tangents at P, Q are at right angles to one another.

29. Given two tangents and their points of contact, determine the focus and vertex.

30. OP, OQ are tangents to the parabola from O, prove
$$OP^2 : OQ^2 :: SP : SQ.$$

31. The lines joining the intersections of the tangents to confocal parabolas drawn to both curves at their points of intersection pass through the common focus.

32. OP, OQ are tangents to the parabola from O; if the chord PQ meets the directrix in F, prove that OSF is a right angle. If OK be at right angles to the directrix, prove SK is at right angles to PQ. If PQ cuts the axis in N, prove that KN is parallel to OS. If OM be drawn to meet the axis at right angles $AM = AN$.

33. OP, OQ are tangents to the parabola from O. Prove that when PQ moves parallel to itself, O moves parallel to the axis: when PQ moves round a point in the axis, O moves at right angles to the axis: when PQ moves round a point in the directrix, O moves in a straight line to S.

34. A line drawn from the focus to meet the tangent at a constant angle, has its point of intersection with it, on one of two fixed tangents.

35. Given one tangent to a parabola, to draw two others which make a given angle with it. In what case is one of the tangents removed to an infinite distance?

36. BC the portion of a tangent intercepted between two other tangents AB and AC is bisected by D the point of contact. Prove that SA is a fourth proportional to AD, AB and AC.

37. The portion of any tangent between tangents that meet on the directrix, subtends a right angle at the focus.

38. The tangent at P meets the directrix in F: from any point O in PF, and from F tangents are drawn to the curve, prove that they meet in the line through S at right angles to OS.

39. The chord PQ is a normal at P and QR is drawn parallel

to the axis to meet PP', the double ordinate through P, produced in R: prove that $PP' \cdot PR$ is constant.

40. The ordinate through the middle point of $NG = PG$.

41. An ellipse and parabola have a common focus and directrix: diagonals of the quadrilateral formed by joining the four points where the tangents at the extremities of the axis major cut the parabola pass through the focus and through the extremities of the axis minor.

42. An hyperbola is confocal with a parabola, and has the tangent at the vertex of the parabola for its nearer directrix. Prove that the tangent to the parabola at the point of intersection passes through the further vertex of the hyperbola.

Miscellaneous Examples.

1. The orthogonal projection of a parabola is a parabola.

2. The projection of a parabolic section of a cone on a plane at right angles to the axis of the cone is a parabola having for its focus the point where the axis cuts the plane on which the projection is made.

3. $CS =$ the part of the generating line of the cone which has the same projection on the axis as CA has: this was proved for the ellipse, extend the proof to the hyperbola.

4. If an elliptic or hyperbolic section of a cone be projected on a plane through one vertex at right angles to the axis of the cone, CS^2 is diminished by the square on the distance of C from the plane of projection, and one focus of the projected curve will lie at the intersection of the axis of the cone with this plane.

5. All parabolas cut by parallel planes from a given cone have their foci on a straight line through the vertex of the cone.

6. Given a right cone and a point within it; only two sections have this point for focus and their planes are equally inclined to the line joining the point to the vertex.

7. Given the vertex of a cone and the centre of a sphere inscribed in it: all sections made by planes at right angles to a generating line and to the plane of the paper containing the centre and vertex, will have one of their foci on a circle which touches the axis of the cone at the centre of the sphere.

8. Two cones touch the same two spheres, prove that by whatever planes the two cones are cut, the ratio of their eccentricities is constant.

9. Two cones have supplementary angles and are placed with their vertices and one generating line of each coincident. Curves are cut from them by a plane at right angles to the coincident generating lines: shew that the directrices of either curve pass through the foci of the other.

10. The intersection of a plane with a cylinder is an ellipse with foci at the points of contact of the plane and two spheres inscribed in the cylinder.

October, 1884.

A Catalogue

OF

Educational Books

PUBLISHED BY

Macmillan & Co.,

BEDFORD STREET, STRAND, LONDON.

INDEX.

CLASSICS— PAGE
  Elementary Classics 3
  Classical Series 6
  Classical Library, (1) Texts, (2) Translations . . . 11
  Grammar, Composition, and Philology 15
  Antiquities, Ancient History, and Philosophy . . 20

MATHEMATICS—
  Arithmetic 22
  Algebra 24
  Euclid, and Elementary Geometry 24
  Mensuration 25
  Higher Mathematics 26

SCIENCE—
  Natural Philosophy 34
  Astronomy 39
  Chemistry 39
  Biology 41
  Medicine 45
  Anthropology 45
  Physical Geography, and Geology 46
  Agriculture 46
  Political Economy 47
  Mental and Moral Philosophy 48

HISTORY AND GEOGRAPHY 49

MODERN LANGUAGES AND LITERATURE—
  English 53
  French 59
  German 62
  Modern Greek 63
  Italian 63

DOMESTIC ECONOMY 63

ART AND KINDRED SUBJECTS 64

WORKS ON TEACHING 65

DIVINITY 65

29 AND 30, BEDFORD STREET, COVENT GARDEN,
LONDON, W.C., *October*, 1884.

CLASSICS.

ELEMENTARY CLASSICS.

18mo, Eighteenpence each.

THIS SERIES FALLS INTO TWO CLASSES—

(1) First Reading Books for Beginners, provided not only with **Introductions and Notes**, but with **Vocabularies**, and in some cases with **Exercises** based upon the Text.

(2) Stepping-stones to the study of particular authors, intended for more advanced students who are beginning to read such authors as Terence, Plato, the Attic Dramatists, and the harder parts of Cicero, Horace, Virgil, and Thucydides.

These are provided with Introductions and Notes, but **no Vocabulary**. The Publishers have been led to provide the more strictly Elementary Books with Vocabularies by the representations of many teachers, who hold that beginners do not understand the use of a Dictionary, and of others who, in the case of middle-class schools where the cost of books is a serious consideration, advocate the Vocabulary system on grounds of economy. It is hoped that the two parts of the Series, fitting into one another, may together fulfil all the requirements of Elementary and Preparatory Schools, and the Lower Forms of Public Schools.

The following Elementary Books, with Introductions, Notes, **and Vocabularies**, and in some cases with **Exercises**, are either ready or in preparation:—

Cæsar.—THE GALLIC WAR. BOOK I. Edited by A. S. WALPOLE, M.A. [*Ready.*

THE INVASION OF BRITAIN. Being Selections from Books IV. and V. of the "De Bello Gallico." Adapted for the use of Beginners. With Notes, Vocabulary, and Exercises, by W. WELCH, M.A., and C. G. DUFFIELD, M.A. [*Ready.*

THE GALLIC WAR. BOOKS II. AND III. Edited by the Rev. W. G. RUTHERFORD, M.A., LL.D., Head-Master of Westminster School. [*Ready.*

THE GALLIC WAR. SCENES FROM BOOKS V. AND VI. Edited by C. COLBECK, M.A., Assistant-Master at Harrow; formerly Fellow of Trinity College, Cambridge. [*Ready.*

Cicero.—DE SENECTUTE. Edited by E. S. SHUCKBURGH, M.A., late Fellow of Emmanuel College, Cambridge.
[*In preparation.*

DE AMICITIA. By the same Editor. [*In preparation.*

STORIES OF ROMAN HISTORY. Adapted for the Use of Beginners. With Notes, Vocabulary, and Exercises, by the Rev. G. E. JEANS, M.A., Fellow of Hertford College, Oxford, and A. V. JONES, M.A., Assistant-Masters at Haileybury College.
[*Ready.*

Eutropius.—Adapted for the Use of Beginners. With Notes, Vocabulary, and Exercises, by WILLIAM WELCH, M.A., and C. G. DUFFIELD, M.A., Assistant-Masters at Surrey County School, Cranleigh. [*Ready.*

Greek Testament.—SELECTIONS FROM THE GOSPELS. Edited by Rev. A. CALVERT, M.A., late Fellow of St. John's College, Cambridge. [*In preparation.*

Homer.—ILIAD. BOOK I. Edited by Rev. JOHN BOND, M.A., and A. S. WALPOLE, M.A. [*Ready.*

ILIAD. BOOK XVIII. THE ARMS OF ACHILLES. Edited by S. R. JAMES, M.A., Assistant-Master at Eton College.
[*Ready.* VOCABULARY *in preparation.*

ODYSSEY. BOOK I. Edited by Rev. JOHN BOND, M.A. and A. S. WALPOLE, M.A. [*Ready.*

CLASSICAL SERIES.

Horace.—ODES. BOOKS I.—IV. Edited by T. E. PAGE, M.A., late Fellow of St. John's College, Cambridge; Assistant-Master at the Charterhouse. Each 1s. 6d.
[*Ready.* VOCABULARY to Book III. *in preparation.*

Livy.—BOOK I. Edited by H. M. STEPHENSON, M.A., Head Master of St. Peter's School, York. [*Ready.*
THE HANNIBALIAN WAR. Being part of the XXI. AND XXII. BOOKS OF LIVY, adapted for the use of beginners, by G. C. MACAULAY, M.A., Assistant-Master at Rugby; formerly Fellow of Trinity College, Cambridge. [*Ready.*
THE SIEGE OF SYRACUSE. Adapted for the Use of Beginners. With Notes, Vocabulary, and Exercises, by GEORGE RICHARDS, M.A., and A. S. WALPOLE, M.A. [*In the press.*

Ovid.—SELECTIONS. Edited by E. S. SHUCKBURGH, M.A., late Fellow and Assistant-Tutor of Emmanuel College, Cambridge.
[*Ready.*

Phædrus.—SELECT FABLES. Adapted for the Use of Beginners. With Notes, Exercises, and Vocabularies, by A. S. WALPOLE, M.A. [*Ready.*

Thucydides.—THE RISE OF THE ATHENIAN EMPIRE. BOOK I. cc. LXXXIX. — CXVII. AND CXXVIII. — CXXXVIII. Edited with Notes, Vocabulary and Exercises, by F. H. COLSON, M.A., Senior Classical Master at Bradford Grammar School; Fellow of St. John's College, Cambridge.
[*Ready.*

Virgil.—ÆNEID. BOOK I. Edited by A. S. WALPOLE, M.A.
[*Ready.*
ÆNEID. BOOK V. Edited by Rev. A. CALVERT, M.A., late Fellow of St. John's College, Cambridge. [*Ready.*
SELECTIONS. Edited by E. S. SHUCKBURGH, M.A.
[*Ready.*

Xenophon.—ANABASIS. BOOK I. Edited by A. S. WALPOLE, M.A. [*Ready.*

The following more advanced Books, with Introductions and Notes, but **no Vocabulary**, are either ready, or in preparation:—

Cicero.—SELECT LETTERS. Edited by Rev. G. E. JEANS, M.A., Fellow of Hertford College, Oxford, and Assistant-Master at Haileybury College. [*Ready.*

Euripides.—HECUBA. Edited by Rev. JOHN BOND, M.A. and A. S. WALPOLE, M.A. [*Ready.*

Herodotus.—SELECTIONS FROM BOOKS VII. AND VIII. THE EXPEDITION OF XERXES. Edited by A. H. COOKE, M.A., Fellow and Lecturer of King's College, Cambridge.
[*Ready.*

Horace. — SELECTIONS FROM THE SATIRES AND EPISTLES. Edited by Rev. W. J. V. BAKER, M.A., Fellow of St. John's College, Cambridge ; Assistant-Master in Marlborough College. [*Ready.*
SELECT EPODES AND ARS POETICA. Edited by H. A. DALTON, M.A., formerly Senior Student of Christchurch ; Assistant-Master in Winchester College. [*Ready.*

Livy.—THE LAST TWO KINGS OF MACEDON. SCENES FROM THE LAST DECADE OF LIVY. Selected and Edited by F. H. RAWLINS, M.A., Fellow of King's College, Cambridge; and Assistant-Master at Eton College. [*In preparation.*

Plato.—EUTHYPHRO AND MENEXENUS. Edited by C. E. GRAVES, M.A., Classical Lecturer and late Fellow of St. John's College, Cambridge. [*Ready.*

Terence.—SCENES FROM THE ANDRIA. Edited by F. W. CORNISH, M.A., Assistant-Master at Eton College. [*Ready.*

The Greek Elegiac Poets.— FROM CALLINUS TO CALLIMACHUS. Selected and Edited by Rev. HERBERT KYNASTON, D.D., Principal of Cheltenham College, and formerly Fellow of St. John's College, Cambridge. [*Ready.*

Thucydides.—BOOK IV. CHS. I.—XLI. THE CAPTURE OF SPHACTERIA. Edited by C. E. GRAVES, M.A. [*Ready.*

Virgil.—GEORGICS. BOOK II. Edited by Rev. J. H. SKRINE, M.A., late Fellow of Merton College, Oxford ; Assistant-Master at Uppingham. [*Ready.*

*** *Other Volumes to follow.*

CLASSICAL SERIES
FOR COLLEGES AND SCHOOLS.
Fcap. 8vo.

Being select portions of Greek and Latin authors, edited with Introductions and Notes, for the use of Middle and Upper forms of Schools, or of candidates for Public Examinations at the Universities and elsewhere.

Æschines.— IN CTESIPHONTEM. Edited by Rev. T. GWATKIN, M.A., late Fellow of St. John's College, Cambridge.
[*In the press.*

CLASSICAL SERIES.

Æschylus.—PERSÆ. Edited by A. O. PRICKARD, M.A., Fellow and Tutor of New College, Oxford. With Map. 3*s.* 6*d.*

Catullus.—SELECT POEMS. Edited by F. P. SIMPSON, B.A., late Scholar of Balliol College, Oxford. New and Revised Edition. 5*s.* The Text of this Edition is carefully adapted to School use.

Cicero.—THE CATILINE ORATIONS. From the German of KARL HALM. Edited, with Additions, by A. S. WILKINS, M.A., Professor of Latin at the Owens College, Manchester. New Edition. 3*s.* 6*d.*

PRO LEGE MANILIA. Edited after HALM by Professor A. S. WILKINS, M.A. 3*s.* 6*d.*

THE SECOND PHILIPPIC ORATION. From the German of KARL HALM. Edited, with Corrections and Additions, by JOHN E. B. MAYOR, Professor of Latin in the University of Cambridge, and Fellow of St. John's College. New Edition, revised. 5*s.*

PRO ROSCIO AMERINO. Edited, after HALM, by E. H. DONKIN, M.A., late Scholar of Lincoln College, Oxford; Assistant-Master at Sherborne School. 4*s.* 6*d.*

PRO P. SESTIO. Edited by Rev. H. A. HOLDEN, M.A., LL.D., late Fellow of Trinity College, Cambridge; and late Classical Examiner to the University of London. 5*s.*

Demosthenes.—DE CORONA. Edited by B. DRAKE, M.A., late Fellow of King's College, Cambridge. New and revised Edition. 4*s.* 6*d.*

ADVERSUS LEPTINEM. Edited by Rev. J. R. KING, M.A., Fellow and Tutor of Oriel College, Oxford. 4*s.* 6*d.*

THE FIRST PHILIPPIC. Edited, after C. REHDANTZ, by Rev. T. GWATKIN, M.A., late Fellow of St. John's College, Cambridge. 2*s.* 6*d.*

Euripides.—BACCHAE. Edited by E. S. SHUCKBURGH, M.A., late Fellow of Emmanuel College, Cambridge. [*In preparation.*

HIPPOLYTUS. Edited by J. P. MAHAFFY, M.A., Fellow and Professor of Ancient History in Trinity College, Dublin, and J. B. BURY, Scholar of Trinity College, Dublin. 3*s.* 6*d.*

Euripides.—MEDEA. Edited by A. W. VERRALL, M.A., Fellow and Lecturer of Trinity College, Cambridge. 3s. 6d.

IPHIGENIA IN TAURIS. Edited by E. B. ENGLAND, M.A., Lecturer at the Owens College, Manchester. 4s. 6d.

Herodotus.—BOOKS V. AND VI. Edited by Rev. A. H. COOKE, M.A., Fellow of King's College, Cambridge. [*In preparation*

BOOKS VII. AND VIII. THE INVASION OF GREECE BY XERXES. Edited by THOMAS CASE, M.A., formerly Fellow of Brasenose College, Oxford. [*In preparation.*

Homer.—ILIAD. BOOKS I., IX., XI., XVI.—XXIV. THE STORY OF ACHILLES. Edited by the late J. H. PRATT, M.A., and WALTER LEAF, M.A., Fellows of Trinity College, Cambridge. 6s.

ODYSSEY. BOOK IX. Edited by Prof. JOHN E. B. MAYOR. 2s. 6d.

ODYSSEY. BOOKS XXI.—XXIV. THE TRIUMPH OF ODYSSEUS. Edited by S. G. HAMILTON, B.A., Fellow of Hertford College, Oxford. 3s. 6d.

Horace.—THE ODES. Edited by T. E. PAGE, M.A., formerly Fellow of St. John's College, Cambridge; Assistant-Master at Charterhouse. 6s. (BOOKS I., II., III., and IV. separately, 2s. each.)

THE SATIRES. Edited by ARTHUR PALMER, M.A., Fellow of Trinity College, Dublin; Professor of Latin in the University of Dublin. 6s.

THE EPISTLES AND ARS POETICA. Edited by Professor A. S. WILKINS, M.A. [*In the press.*

Juvenal. THIRTEEN SATIRES. Edited, for the Use of Schools, by E. G. HARDY, M.A., Head-Master of Grantham Grammar School; late Fellow of Jesus College, Oxford. 5s.
The Text of this Edition is carefully adapted to School use.

SELECT SATIRES. Edited by Professor JOHN E. B. MAYOR. X. AND XI. 3s. 6d. XII.—XVI. 4s. 6d.

Livy.—BOOKS II. AND III. Edited by Rev. H. M. STEPHENSON, M.A., Head-Master of St. Peter's School, York. 5s.

BOOKS XXI. AND XXII. Edited by the Rev. W. W. CAPES, M.A., Reader in Ancient History at Oxford. With Maps. 5s.

BOOKS XXIII AND XXIV. Edited by G. C. MACAULAY, M.A., Assistant-Master at Rugby. [*In preparation.*

CLASSICAL SERIES. 9

Lucretius. BOOKS I.—III. Edited by J. H. WARBURTON LEE, M.A., late Scholar of Corpus Christi College, Oxford, and Assistant-Master at Rossall. [*In preparation.*

Lysias.—SELECT ORATIONS. Edited by E. S. SHUCKBURGH, M.A., Assistant-Master at Eton College. 6*s.*

Martial. — SELECT EPIGRAMS. Edited by Rev. H. M. STEPHENSON, M.A. 6*s.*

Ovid.—FASTI. Edited by G. H. HALLAM, M.A., Fellow of St. John's College, Cambridge, and Assistant-Master at Harrow. With Maps. 5*s.*
HEROIDUM EPISTULÆ XIII. Edited by E. S. SHUCKBURGH, M.A. 4*s.* 6*d.*
METAMORPHOSES. BOOKS XIII. AND XIV. Edited by C. SIMMONS, M.A. [*In the press.*

Plato.—MENO. Edited by E. S. THOMPSON, M.A., Fellow of Christ's College, Cambridge. [*In preparation.*
APOLOGY AND CRITO. Edited by F. J. H. JENKINSON, M.A., Fellow of Trinity College, Cambridge. [*In preparation.*
THE REPUBLIC. BOOKS I.—V. Edited by T. H. WARREN, M.A., Fellow of Magdalen College, Oxford. [*In preparation.*

Plautus.—MILES GLORIOSUS. Edited by R. Y. TYRRELL, M.A., Fellow and Professor of Greek in Trinity College, Dublin. 5*s.*

Pliny.—LETTERS. BOOK III. Edited by Professor JOHN E. B. MAYOR. With Life of Pliny by G. H. RENDALL, M.A. 5*s.*

Plutarch.—LIFE OF THEMISTOKLES. Edited by Rev. H. A. HOLDEN, M.A., LL.D. 5*s.*

Propertius.—SELECT POEMS. Edited by Professor J. P. POSTGATE, M.A., Fellow of Trinity College, Cambridge. 6*s.*

Sallust.—CATILINA AND JUGURTHA. Edited by C. MERIVALE, D.D., Dean of Ely. New Edition, carefully revised and enlarged, 4*s.* 6*d.* Or separately, 2*s.* 6*d.* each.
BELLUM CATULINAE. Edited by A. M. COOK, M.A., Assistant Master at St. Paul's School. 4*s.* 6*d.*

Sophocles.—ANTIGONE. Edited by Rev. JOHN BOND, M.A., and A. S. WALPOLE, M.A. [*In preparation.*

Tacitus.—AGRICOLA AND GERMANIA. Edited by A. J. CHURCH, M.A., and W. J. BRODRIBB, M.A., Translators of Tacitus. New Edition, 3s. 6d. Or separately, 2s. each.

THE ANNALS. BOOK VI. By the same Editors. 2s. 6d.

THE HISTORY. BOOKS I. AND II. Edited by A. D. GODLEY, M.A. [*In preparation.*

THE ANNALS. BOOKS I. AND II. Edited by J. S. REID, M.A. [*In preparation.*

Terence.—HAUTON TIMORUMENOS. Edited by E. S. SHUCKBURGH, M.A. 3s. With Translation, 4s. 6d.

PHORMIO. Edited by Rev. JOHN BOND, M.A., and A. S. WALPOLE, B.A. 4s. 6d.

Thucydides. BOOK IV. Edited by C. E. GRAVES, M.A., Classical Lecturer, and late Fellow of St. John's College, Cambridge. 5s.

BOOKS I. II. III. AND V. By the same Editor. To be published separately. [*In preparation.*

BOOKS VI. AND VII. THE SICILIAN EXPEDITION. Edited by the Rev. PERCIVAL FROST, M.A., late Fellow of St. John's College, Cambridge. New Edition, revised and enlarged, with Map. 5s.

Virgil.—ÆNEID. BOOKS II. AND III. THE NARRATIVE OF ÆNEAS. Edited by E. W. HOWSON, M.A., Fellow of King's College, Cambridge, and Assistant-Master at Harrow. 3s.

Xenophon.—HELLENICA, BOOKS I. AND II. Edited by H. HAILSTONE, B.A., late Scholar of Peterhouse, Cambridge. With Map. 4s. 6d.

CYROPÆDIA. BOOKS VII. AND VIII. Edited by ALFRED GOODWIN, M.A., Professor of Greek in University College, London. 5s.

MEMORABILIA SOCRATIS. Edited by A. R. CLUER, B.A. Balliol College, Oxford. 6s.

THE ANABASIS. BOOKS I.—IV. Edited by Professors W. W. GOODWIN and J. W. WHITE. Adapted to Goodwin's Greek Grammar. With a Map. 5s.

Xenophon.—HIERO. Edited by Rev. H. A. HOLDEN, M.A., LL.D. 3s. 6d.
OECONOMICUS. By the same Editor. With Introduction, Explanatory Notes, Critical Appendix, and Lexicon. 6s.
₊ *Other Volumes will follow.*

CLASSICAL LIBRARY.

(1) **Texts**, Edited with **Introductions and Notes**, for the use of Advanced Students. (2) **Commentaries and Translations.**

Æschylus.—THE EUMENIDES. The Greek Text, with Introduction, English Notes, and Verse Translation. By BERNARD DRAKE, M.A., late Fellow of King's College, Cambridge. 8vo. 5s.

AGAMEMNON, CHOEPHORŒ, AND EUMENIDES. Edited, with Introduction and Notes, by A. O. PRICKARD, M.A., Fellow and Tutor of New College, Oxford. 8vo. [*In preparation.*

AGAMEMNO. Emendavit DAVID S. MARGOLIOUTH, Coll. Nov. Oxon. Soc. Demy 8vo. 2s. 6d.

Antoninus, Marcus Aurelius.—BOOK IV. OF THE MEDITATIONS. The Text Revised, with Translation and Notes, by HASTINGS CROSSLEY, M.A., Professor of Greek in Queen's College, Belfast. 8vo. 6s.

Aristotle.—THE METAPHYSICS. BOOK I. Translated by a Cambridge Graduate. 8vo. 5s. [*Book II. in preparation.*

THE POLITICS. Edited, after SUSEMIHL, by R. D. HICKS, M.A., Fellow of Trinity College, Cambridge. 8vo. [*In the press.*

THE POLITICS. Translated by Rev. J. E. C. WELLDON, M.A., Fellow of King's College, Cambridge, and Master of Dulwich College. Crown 8vo. 10s. 6d.

THE RHETORIC. By the same Translator. [*In preparation.*

AN INTRODUCTION TO ARISTOTLE'S RHETORIC. With Analysis, Notes, and Appendices. By E. M. COPE, Fellow and Tutor of Trinity College, Cambridge. 8vo. 14s.

THE SOPHISTICI ELENCHI. With Translation and Notes by E. POSTE, M.A., Fellow of Oriel College, Oxford. 8vo. 8s. 6d.

Aristophanes.—THE BIRDS. Translated into English Verse, with Introduction, Notes, and Appendices, by B. H. KENNEDY, D.D., Regius Professor of Greek in the University of Cambridge. Crown 8vo. 6s. Help Notes to the same, for the use of Students, 1s. 6d.

Attic Orators.—FROM ANTIPHON TO ISAEOS. By R. C. JEBB, M.A., LL.D., Professor of Greek in the University of Glasgow. 2 vols. 8vo. 25s.

SELECTIONS FROM ANTIPHON, ANDOKIDES, LYSIAS, ISOKRATES, AND ISÆOS. Edited, with Notes, by Professor JEBB. Being a companion volume to the preceding work. 8vo. 12s. 6d.

Babrius. Edited, with Introductory Dissertations, Critical Notes, Commentary and Lexicon. By Rev. W. GUNION RUTHERFORD, M.A., LL.D., Head-Master of Westminster School. 8vo. 12s. 6d.

Cicero.—THE ACADEMICA. The Text revised and explained by J. S. REID, M.L., Fellow of Caius College, Cambridge. New Edition. With Translation. 8vo. [*In the press.*

THE ACADEMICS. Translated by J. S. REID, M.L. 8vo. 5s. 6d.

SELECT LETTERS. After the Edition of ALBERT WATSON, M.A. Translated by G. E. JEANS, M.A., Fellow of Hertford College, Oxford, and Assistant-Master at Haileybury. 8vo. 10s. 6d.

(See also *Classical Series*.)

Euripides.—MEDEA. Edited, with Introduction and Notes, by A. W. VERRALL, M.A., Fellow and Lecturer of Trinity College, Cambridge. 8vo. 7s. 6d.

INTRODUCTION TO THE STUDY OF EURIPIDES. By Professor J. P. MAHAFFY. Fcap. 8vo. 1s. 6d. (*Classical Writers Series.*)

(See also *Classical Series*.)

Herodotus.—BOOKS I.—III. THE ANCIENT EMPIRES OF THE EAST. Edited, with Notes, Introductions, and Appendices, by A. H. SAYCE, Deputy-Professor of Comparative Philology, Oxford; Honorary LL.D., Dublin. Demy 8vo. 16s.

BOOKS IV.—IX. Edited by REGINALD W. MACAN, M.A., Lecturer in Ancient History at Brasenose College, Oxford. 8vo. [*In preparation.*

CLASSICAL LIBRARY. 13

Homer.—THE ILIAD. Edited, with Introduction and Notes, by WALTER LEAF, M.A., Fellow of Trinity College, Cambridge, and the late J. H. PRATT, M.A. 8vo. [*In preparation.*
THE ILIAD. Translated into English Prose. By ANDREW LANG, M.A., WALTER LEAF, M.A., and ERNEST MYERS, M.A. Crown 8vo. 12s. 6d.
THE ODYSSEY. Done into English by S. H. BUTCHER, M.A., Professor of Greek in the University of Edinburgh, and ANDREW LANG, M.A., late Fellow of Merton College, Oxford. Fourth Edition, revised and corrected. Crown 8vo. 10s. 6d.
INTRODUCTION TO THE STUDY OF HOMER. By the Right Hon. W. E. GLADSTONE, M.P. 18mo. 1s. (*Literature Primers.*)
HOMERIC DICTIONARY. For Use in Schools and Colleges. Translated from the German of Dr. G. AUTENRIETH, with Additions and Corrections, by R. P. KEEP, Ph.D. With numerous Illustrations. Crown 8vo. 6s.
(See also *Classical Series.*)

Horace.—THE WORKS OF HORACE RENDERED INTO ENGLISH PROSE. With Introductions, Running Analysis, Notes, &c. By J. LONSDALE, M.A., and S. LEE, M.A. (*Globe Edition.*) 3s. 6d.
ESSAYS UPON. By A. W. VERRAL, M.A. 8vo. [*In the press.*
(See also *Classical Series.*)

Juvenal.—THIRTEEN SATIRES OF JUVENAL. With a Commentary. By JOHN E. B. MAYOR, M.A., Professor of Latin in the University of Cambridge. Second Edition, enlarged. Crown 8vo. Vol. I. 7s. 6d. Vol. II. 10s. 6d.
THIRTEEN SATIRES. Translated into English after the Text of J. E. B. MAYOR by HERBERT STRONG, M.A., Professor of Latin, and ALEXANDER LEEPER, M.A., Warden of Trinity College, in the University of Melbourne. Crown 8vo. 3s. 6d.
(See also *Classical Series.*)

Livy. BOOKS XXI.—XXV. Translated by ALFRED JOHN CHURCH, M.A., of Lincoln College, Oxford, Professor of Latin, University College, London, and WILLIAM JACKSON BRODRIBB, M.A., late Fellow of St. John's College, Cambridge. Crown 8vo 7s. 6d.
INTRODUCTION TO THE STUDY OF LIVY. By Rev. W. W. CAPES, Reader in Ancient History at Oxford. Fcap. 8vo. 1s. 6d. (*Classical Writers Series.*)
(See also *Classical Series.*)

Martial.—BOOKS I. AND II. OF THE EPIGRAMS. Edited, with Introduction and Notes, by Professor J. E. B. MAYOR, M.A. 8vo. [*In the press.*
(See also *Classical Series.*)

Pausanias.—DESCRIPTION OF GREECE. Translated by J. G. FRAZER, M.A., Fellow of Trinity College, Cambridge.
[*In preparation.*

Phrynichus.—THE NEW PHRYNICHUS; being a Revised Text of the Ecloga of the Grammarian Phrynichus. With Introduction and Commentary by Rev. W. GUNION RUTHERFORD, M.A., LL.D., Head Master of Westminster School. 8vo. 18s.

Pindar.—THE EXTANT ODES OF PINDAR. Translated into English, with an Introduction and short Notes, by ERNEST MYERS, M.A., late Fellow of Wadham College, Oxford. Second Edition. Crown 8vo. 5s.

Plato.—PHÆDO. Edited, with Introduction, Notes, and Appendices, by R. D. ARCHER-HIND, M.A., Fellow of Trinity College, Cambridge. 8vo. 8s. 6d.

PHILEBUS. Edited, with Introduction and Notes, by HENRY JACKSON, M.A., Fellow of Trinity College, Cambridge. 8vo.
[*In preparation.*

THE REPUBLIC OF PLATO. Translated into English, with an Analysis and Notes, by J. LL. DAVIES, M.A., and D. J. VAUGHAN, M.A. 18mo. 4s. 6d.

EUTHYPHRO, APOLOGY, CRITO, AND PHÆDO. Translated by F. J. CHURCH. Crown 8vo. 4s. 6d.
(See also *Classical Series.*)

Plautus.—THE MOSTELLARIA OF PLAUTUS. With Notes, Prolegomena, and Excursus. By WILLIAM RAMSAY, M.A., formerly Professor of Humanity in the University of Glasgow. Edited by Professor GEORGE G. RAMSAY, M.A., of the University of Glasgow. 8vo. 14s.
(See also *Classical Series.*)

Sallust.—CATILINE AND JUGURTHA. Translated, with Introductory Essays, by A. W. POLLARD, B.A. Crown 8vo. 6s.
(See also *Classical Series.*)

Studia Scenica.—Part I., Section I. Introductory Study on the Text of the Greek Dramas. The Text of SOPHOCLES' TRACHINIAE, 1-300. By DAVID S. MARGOLIOUTH, Fellow of New College, Oxford. Demy 8vo. 2s. 6d.

Tacitus.—THE ANNALS. Edited, with Introductions and Notes, by G. O. HOLBROOKE, M.A., Professor of Latin in Trinity College, Hartford, U.S.A. With Maps. 8vo. 16s.

THE ANNALS. Translated by A. J. CHURCH, M.A., Professor of Latin in the University of London, and W. J. BRODRIBB, M.A. With Notes and Maps. New Edition. Crown 8vo. 7s. 6d.

THE HISTORY. Edited, with Introduction and Notes, by Rev. WALTER SHORT, M.A., and Rev. W. A. SPOONER, M.A. Fellows of New College, Oxford. 8vo. [*In preparation.*

THE HISTORY. Translated by A. J. CHURCH, M.A., Professor of Latin in the University of London, and W. J. BRODRIBB, M.A. With Notes and a Map. New Edition. Crown 8vo. 6s.

THE AGRICOLA AND GERMANY, WITH THE DIALOGUE ON ORATORY. Translated by A. J. CHURCH, M.A., and W. J. BRODRIBB, M.A. With Notes and Maps. New and Revised Edition. Crown 8vo. 4s. 6d.

INTRODUCTION TO THE STUDY OF TACITUS. By A. J. CHURCH, M.A. and W. J. BRODRIBB, M.A. Fcap. 8vo. 18mo. 1s. 6d. (*Classical Writers Series.*)

Theocritus, Bion, and Moschus. Rendered into English Prose with Introductory Essay by ANDREW LANG, M.A. Crown 8vo. 6s.

Virgil.—THE WORKS OF VIRGIL RENDERED INTO ENGLISH PROSE, with Notes, Introductions, Running Analysis, and an Index, by JAMES LONSDALE, M.A., and SAMUEL LEE, M.A. New Edition. Globe 8vo. 3s. 6d.

THE ÆNEID. Translated by J. W. MACKAIL, M.A., Fellow of Balliol College, Oxford. Crown 8vo. [*In the press.*

GRAMMAR, COMPOSITION, & PHILOLOGY.

Belcher.—SHORT EXERCISES IN LATIN PROSE COMPOSITION AND EXAMINATION PAPERS IN LATIN GRAMMAR, to which is prefixed a Chapter on Analysis of Sentences. By the Rev. H. BELCHER, M.A., Assistant-Master in King's College School, London. New Edition. 18mo. 1s. 6d.

KEY TO THE ABOVE (for Teachers only). 2s. 6d.

SHORT EXERCISES IN LATIN PROSE COMPOSITION Part II., On the Syntax of Sentences, with an Appendix, including EXERCISES IN LATIN IDIOMS, &c. 18mo. 2s.

KEY TO THE ABOVE (for Teachers only). 3s.

Blackie.—GREEK AND ENGLISH DIALOGUES FOR USE IN SCHOOLS AND COLLEGES. By JOHN STUART BLACKIE, Emeritus Professor of Greek in the University of Edinburgh. New Edition. Fcap. 8vo. 2s. 6d.

Bryans.—LATIN PROSE EXERCISES BASED UPON CAESAR'S GALLIC WAR. With a Classification of Cæsar's Chief Phrases and Grammatical Notes in Cæsar's Usages. By CLEMENT BRYANS, M.A., Assistant-Master in Dulwich College, late Scholar in King's College, Cambridge, and Bell University Scholar. Extra fcap. 8vo. 2s. 6d.

GREEK PROSE EXERCISES based upon Thucydides. By the same Author. Extra fcap. 8vo. [*In preparation.*

Colson.—A FIRST GREEK READER. By F. H. COLSON, M.A., Fellow of St. John's College, Cambridge, and Senior Classical Master at Bradford Grammar School. Globe 8vo.
[*In preparation.*

Eicke.—FIRST LESSONS IN LATIN. By K. M. EICKE, B.A., Assistant-Master in Oundle School. Globe 8vo. 2s.

Ellis.—PRACTICAL HINTS ON THE QUANTITATIVE PRONUNCIATION OF LATIN, for the use of Classical Teachers and Linguists. By A. J. ELLIS, B.A., F.R.S. Extra fcap. 8vo. 4s. 6d.

England.—EXERCISES ON LATIN SYNTAX AND IDIOM, ARRANGED WITH REFERENCE TO ROBY'S SCHOOL LATIN GRAMMAR. By E. B. ENGLAND, M.A., Assistant Lecturer at the Owens College, Manchester. Crown 8vo. 2s. 6d. Key for Teachers only, 2s. 6d.

Goodwin.—Works by W. W. GOODWIN, LL.D., Professor of Greek in Harvard University, U.S.A.

SYNTAX OF THE MOODS AND TENSES OF THE GREEK VERB. New Edition, revised. Crown 8vo. 6s. 6d.

A GREEK GRAMMAR. New Edition, revised. Crown 8vo. 6s.
"It is the best Greek Grammar of its size in the English language."— ATHENÆUM.

A GREEK GRAMMAR FOR SCHOOLS. Crown 8vo. 3s. 6d.

Greenwood.—THE ELEMENTS OF GREEK GRAMMAR, including Accidence, Irregular Verbs, and Principles of Derivation and Composition; adapted to the System of Crude Forms. By J. G. GREENWOOD, Principal of Owens College, Manchester. New Edition. Crown 8vo. 5s. 6d.

CLASSICAL PUBLICATIONS.

Hadley and Allen.—A GREEK GRAMMAR. By the late Prof. HADLEY. New Edition revised by Prof. ALLEN, of Harvard College. Crown 8vo. [*In the press.*

Hodgson.—MYTHOLOGY FOR LATIN VERSIFICATION. A brief Sketch of the Fables of the Ancients, prepared to be rendered into Latin Verse for Schools. By F. HODGSON, B.D., late Provost of Eton. New Edition, revised by F. C. HODGSON, M.A. 18mo. 3s.

Jackson.—FIRST STEPS TO GREEK PROSE COMPOSITION. By BLOMFIELD JACKSON, M.A., Assistant-Master in King's College School, London. New Edition, revised and enlarged. 18mo. 1s. 6d.

KEY TO FIRST STEPS (for Teachers only). 18mo. 3s. 6d.

SECOND STEPS TO GREEK PROSE COMPOSITION, with Miscellaneous Idioms, Aids to Accentuation, and Examination Papers in Greek Scholarship. 18mo. 2s. 6d.

KEY TO SECOND STEPS (for Teachers only). 18mo. 3s. 6d.

Kynaston.—EXERCISES IN THE COMPOSITION OF GREEK IAMBIC VERSE by Translations from English Dramatists. By Rev. H. KYNASTON, D.D., Principal of Cheltenham College. With Introduction, Vocabulary, &c. Extra fcap. 8vo. 4s. 6d.

KEY TO THE SAME (for Teachers only). Extra fcap. 8vo. 4s. 6d.

Lupton.—ELEMENTARY EXERCISES IN LATIN VERSE COMPOSITION. By Rev. J. H. LUPTON, M.A., Sur-Master in St. Paul's School. Globe 8vo. [*In preparation.*

Macmillan.—FIRST LATIN GRAMMAR. By M. C. MACMILLAN, M.A., late Scholar of Christ's College, Cambridge; sometime Assistant-Master in St. Paul's School. New Edition, enlarged. 18mo. 1s. 6d. A SHORT SYNTAX is in preparation to follow the ACCIDENCE.

Macmillan's Progressive Latin Course. By A. M. COOK, M.A., Assist. Master at St. Paul's School. [*In preparation.*

Marshall.—A TABLE OF IRREGULAR GREEK VERBS, classified according to the arrangement of Curtius's Greek Grammar. By J. M. MARSHALL, M.A., one of the Masters in Clifton College. 8vo, cloth. New Edition. 1s.

Mayor (John E. B.)—FIRST GREEK READER. Edited after KARL HALM, with Corrections and large Additions by Professor JOHN E. B. MAYOR, M.A., Fellow of St. John's College, Cambridge. New Edition, revised. Fcap. 8vo. 4s. 6d.

Mayor (Joseph B.)—GREEK FOR BEGINNERS. By the Rev. J. B. MAYOR, M.A., Professor of Classical Literature in King's College, London. Part I., with Vocabulary,' 1s. 6d. Parts II. and III., with Vocabulary and Index, 3s. 6d. Complete in one Vol. fcap. 8vo. 4s. 6d.

Nixon.—PARALLEL EXTRACTS arranged for translation into English and Latin, with Notes on Idioms. By J. E. NIXON, M.A., Fellow and Classical Lecturer, King's College, Cambridge. Part I.—Historical and Epistolary. New Edition, revised and enlarged. Crown 8vo. 3s. 6d.

Peile.—A PRIMER OF PHILOLOGY. By J. PEILE, M.A., Fellow and Tutor of Christ's College, Cambridge. 18mo. 1s.

Postgate and Vince.—A DICTIONARY OF LATIN ETYMOLOGY. By J. P. POSTGATE, M.A., and C. A. VINCE, M.A. [*In preparation.*

Potts (A. W.)—Works by ALEXANDER W. POTTS, M.A., I.L.D., late Fellow of St. John's College, Cambridge; Head Master of the Fettes College, Edinburgh.

HINTS TOWARDS LATIN PROSE COMPOSITION. New Edition. Extra fcap. 8vo. 3s.

PASSAGES FOR TRANSLATION INTO LATIN PROSE. Edited with Notes and References to the above. New Edition. Extra fcap. 8vo. 2s. 6d.

LATIN VERSIONS OF PASSAGES FOR TRANSLATION INTO LATIN PROSE (for Teachers only). 2s. 6d.

Reid.—A GRAMMAR OF TACITUS. By J. S. REID, M.L., Fellow of Caius College, Cambridge. [*In preparation.*
A GRAMMAR OF VERGIL. By the same Author.
[*In preparation.*

*** *Similar Grammars to other Classical Authors will probably follow.*

Roby.—A GRAMMAR OF THE LATIN LANGUAGE, from Plautus to Suetonius. By H. J. ROBY, M.A., late Fellow of St. John's College, Cambridge. In Two Parts. Third Edition. Part I. containing:—Book I. Sounds. Book II. Inflexions. Book III. Word-formation. Appendices. Crown 8vo. 8s. 6d. Part II. Syntax, Prepositions, &c. Crown 8vo. 10s. 6d.

" Marked by the clear and practised insight of a master in his art. A book that would do honour to any country."—ATHENÆUM.

CLASSICAL PUBLICATIONS.   19

Roby (*continued*)—
SCHOOL LATIN GRAMMAR. By the same Author. Crown 8vo. 5s.

Rush.—SYNTHETIC LATIN DELECTUS. A First Latin Construing Book arranged on the Principles of Grammatical Analysis. With Notes and Vocabulary. By E. RUSH, B.A. With Preface by the Rev. W. F. MOULTON, M.A., D.D. New and Enlarged Edition. Extra fcap. 8vo. 2s. 6d.

Rust.—FIRST STEPS TO LATIN PROSE COMPOSITION. By the Rev. G. RUST, M.A., of Pembroke College, Oxford, Master of the Lower School, King's College, London. New Edition. 18mo. 1s. 6d.

Rutherford.—Works by the Rev. W. GUNION RUTHERFORD, M.A., I.L.D., Head-Master of Westminster School.
A FIRST GREEK GRAMMAR. New Edition, enlarged. Extra fcap. 8vo. 1s. 6d.
THE NEW PHRYNICHUS; being a Revised Text of the Ecloga of the Grammarian Phrynichus. With Introduction and Commentary. 8vo. 18s.

Simpson.—LATIN PROSE AFTER THE BEST AUTHORS. By F. P. SIMPSON, B.A., late Scholar of Balliol College, Oxford.
I. CÆSAR. Extra fcap. 8vo. [*In the press.*

Thring.—Works by the Rev. E. THRING, M.A., Head-Master of Uppingham School.
A LATIN GRADUAL. A First Latin Construing Book for Beginners. New Edition, enlarged, with Coloured Sentence Maps. Fcap. 8vo. 2s. 6d.
A MANUAL OF MOOD CONSTRUCTIONS. Fcap. 8vo. 1s. 6d.

White.—FIRST LESSONS IN GREEK. Adapted to GOODWIN'S GREEK GRAMMAR, and designed as an introduction to the ANABASIS OF XENOPHON. By JOHN WILLIAMS WHITE, Ph.D., Assistant-Professor of Greek in Harvard University. Crown 8vo. 4s. 6d.

Wright.—Works by J. WRIGHT, M.A., late Head Master of Sutton Coldfield School.
A HELP TO LATIN GRAMMAR; or, The Form and Use of Words in Latin, with Progressive Exercises. Crown 8vo. 4s. 6d.

C 2

Wright (*continued*)—
THE SEVEN KINGS OF ROME. An Easy Narrative, abridged from the First Book of Livy by the omission of Difficult Passages; being a First Latin Reading Book, with Grammatical Notes and Vocabulary. New and revised Edition. Fcap. 8vo. 3s. 6d.
FIRST LATIN STEPS; OR, AN INTRODUCTION BY A SERIES OF EXAMPLES TO THE STUDY OF THE LATIN LANGUAGE. Crown 8vo. 3s.
ATTIC PRIMER. Arranged for the Use of Beginners. Extra fcap. 8vo. 2s. 6d.
A COMPLETE LATIN COURSE, comprising Rules with Examples, Exercises, both Latin and English, on each Rule, and Vocabularies. Crown 8vo. 2s. 6d.

Wright (H. C.)—EXERCISES ON THE LATIN SYNTAX. By H. C. WRIGHT, B.A., Assistant-Master at Haileybury College. 18mo. [*In preparation.*

ANTIQUITIES, ANCIENT HISTORY, AND PHILOSOPHY.

Arnold.—Works by W. T. ARNOLD, B.A.
A HANDBOOK OF LATIN EPIGRAPHY. [*In preparation.*
THE ROMAN SYSTEM OF PROVINCIAL ADMINISTRATION TO THE ACCESSION OF CONSTANTINE THE GREAT. Crown 8vo. 6s.

Beesly.—STORIES FROM THE HISTORY OF ROME. By Mrs. BEESLY. Fcap. 8vo. 2s. 6d.

Classical Writers.—Edited by JOHN RICHARD GREEN, M.A., LL.D. Fcap. 8vo. 1s. 6d. each.
EURIPIDES. By Professor MAHAFFY.
MILTON. By the Rev. STOPFORD A. BROOKE, M.A.
LIVY. By the Rev. W. W. CAPES, M.A.
VIRGIL. By Professor NETTLESHIP, M.A.
SOPHOCLES. By Professor L. CAMPBELL, M.A.
DEMOSTHENES. By Professor S. H. BUTCHER, M.A.
TACITUS. By Professor A. J. CHURCH, M.A., and W. J. BRODRIBB, M.A.

Freeman.—HISTORY OF ROME. By EDWARD A. FREEMAN, D.C.L., LL.D., Hon. Fellow of Trinity College, Oxford, Regius Professor of Modern History in the University of Oxford. (*Historical Course for Schools.*) 18mo. [*In preparation.*
A SCHOOL HISTORY OF ROME. By the same Author. Crown 8vo. [*In preparation.*
HISTORICAL ESSAYS. Second Series. [Greek and Roman History.] By the same Author. 8vo. 10s. 6d.

Fyffe.—A SCHOOL HISTORY OF GREECE. By C. A. FYFFE, M.A., late Fellow of University College, Oxford. Crown 8vo. [*In preparation.*

Geddes. — THE PROBLEM OF THE HOMERIC POEMS. By W. D. GEDDES, Professor of Greek in the University of Aberdeen. 8vo. 14s.

Gladstone.—Works by the Rt. Hon. W. E. GLADSTONE, M.P.
THE TIME AND PLACE OF HOMER. Crown 8vo. 6s. 6d.
A PRIMER OF HOMER. 18mo. 1s.

Jackson.—A MANUAL OF GREEK PHILOSOPHY. By HENRY JACKSON, M.A., Fellow and Prælector in Ancient Philosophy, Trinity College, Cambridge. [*In preparation.*

Jebb.—Works by R. C. JEBB, M.A., Professor of Greek in the University of Glasgow.
THE ATTIC ORATORS FROM ANTIPHON TO ISAEOS. 2 vols. 8vo. 25s.
SELECTIONS FROM THE ATTIC ORATORS, ANTIPHON, ANDOKIDES, LYSIAS, ISOKRATES, AND ISÆOS. Edited, with Notes. Being a companion volume to the preceding work. 8vo. 12s. 6d.
A PRIMER OF GREEK LITERATURE. 18mo. 1s.

Kiepert.—MANUAL OF ANCIENT GEOGRAPHY, Translated from the German of Dr. HEINRICH KIEPERT. Crown 8vo. 5s.

Mahaffy.—Works by J. P. MAHAFFY, M.A., Professor of Ancient History in Trinity College, Dublin, and Hon. Fellow of Queen's College, Oxford.
SOCIAL LIFE IN GREECE; from Homer to Menander. Fourth Edition, revised and enlarged. Crown 8vo. 9s.
RAMBLES AND STUDIES IN GREECE. With Illustrations. Second Edition. With Map. Crown 8vo. 10s. 6d.
A PRIMER OF GREEK ANTIQUITIES. With Illustrations. 18mo. 1s.
EURIPIDES. 18mo. 1s. 6d. (*Classical Writers Series.*)

Mayor (J. E. B.)—BIBLIOGRAPHICAL CLUE TO LATIN LITERATURE. Edited after HÜBNER, with large Additions by Professor JOHN E. B. MAYOR. Crown 8vo. 10s. 6d.

Newton.—ESSAYS IN ART AND ARCHÆOLOGY. By C. T. NEWTON, C.B., D.C.L., Professor of Archæology in University College, London, and Keeper of Greek and Roman Antiquities at the British Museum. 8vo. 12s. 6d.

Ramsay.—A SCHOOL HISTORY OF ROME. By G. G. RAMSAY, M.A., Professor of Humanity in the University of Glasgow. With Maps. Crown 8vo. [*In preparation.*

Schwegler.—A TEXT-BOOK OF GREEK PHILOSOPHY. Translated from the German by HENRY NORMAN. 8vo.
[*In preparation.*

Wilkins.—A PRIMER OF ROMAN ANTIQUITIES. By Professor WILKINS. Illustrated. 18mo. 1s.

MATHEMATICS.

(1) Arithmetic, (2) Algebra, (3) Euclid and Elementary Geometry, (4) Mensuration, (5) Higher Mathematics.

ARITHMETIC.

Aldis.—THE GIANT ARITHMOS. A most Elementary Arithmetic for Children. By MARY STEADMAN ALDIS. With Illustrations. Globe 8vo. 2s. 6d.

Brook-Smith (J.).—ARITHMETIC IN THEORY AND PRACTICE. By J. BROOK-SMITH, M.A., LL.B., St. John's College, Cambridge; Barrister-at-Law; one of the Masters of Cheltenham College. New Edition, revised. Crown 8vo. 4s. 6d.

Candler.—HELP TO ARITHMETIC. Designed for the use of Schools. By H. CANDLER, M.A., Mathematical Master of Uppingham School. Extra fcap. 8vo. 2s. 6d.

Dalton.—RULES AND EXAMPLES IN ARITHMETIC. By the Rev. T. DALTON, M.A., Assistant-Master of Eton College. New Edition. 18mo. 2s. 6d.
[*Answers to the Examples are appended.*

Pedley.—EXERCISES IN ARITHMETIC for the Use of Schools. Containing more than 7,000 original Examples. By S. PEDLEY, late of Tamworth Grammar School. Crown 8vo. 5s.

MATHEMATICS. 23

Smith.—Works by the Rev. BARNARD SMITH, M.A., late Rector of Glaston, Rutland, and Fellow and Senior Bursar of S. Peter's College, Cambridge.

ARITHMETIC AND ALGEBRA, in their Principles and Application; with numerous systematically arranged Examples taken from the Cambridge Examination Papers, with especial reference to the Ordinary Examination for the B.A. Degree. New Edition, carefully Revised. Crown 8vo. 10s. 6d.

ARITHMETIC FOR SCHOOLS. New Edition. Crown 8vo. 4s. 6d.

A KEY TO THE ARITHMETIC FOR SCHOOLS. New Edition. Crown 8vo. 8s. 6d.

EXERCISES IN ARITHMETIC. Crown 8vo, limp cloth, 2s. With Answers, 2s. 6d.

Answers separately, 6d.

SCHOOL CLASS-BOOK OF ARITHMETIC. 18mo, cloth. 3s.
Or sold separately, in Three Parts, 1s. each.

KEYS TO SCHOOL CLASS-BOOK OF ARITHMETIC. Parts I., II., and III., 2s. 6d. each.

SHILLING BOOK OF ARITHMETIC FOR NATIONAL AND ELEMENTARY SCHOOLS. 18mo, cloth. Or separately, Part I. 2d.; Part II. 3d.; Part III. 7d. Answers, 6d.

THE SAME, with Answers complete. 18mo, cloth. 1s. 6d.

KEY TO SHILLING BOOK OF ARITHMETIC. 18mo. 4s. 6d.

EXAMINATION PAPERS IN ARITHMETIC. 18mo. 1s. 6d. The same, with Answers, 18mo, 2s. Answers, 6d.

KEY TO EXAMINATION PAPERS IN ARITHMETIC. 18mo. 4s. 6d.

THE METRIC SYSTEM OF ARITHMETIC, ITS PRINCIPLES AND APPLICATIONS, with numerous Examples, written expressly for Standard V. in National Schools. New Edition. 18mo, cloth, sewed. 3d.

A CHART OF THE METRIC SYSTEM, on a Sheet, size 42 in. by 34 in. on Roller, mounted and varnished. New Edition. Price 3s. 6d.

Also a Small Chart on a Card, price 1d.

EASY LESSONS IN ARITHMETIC, combining Exercises in Reading, Writing, Spelling, and Dictation. Part I. for Standard I. in National Schools. Crown 8vo. 9d.

Smith.—Works by the Rev. BARNARD SMITH, M.A. (*continued*)—
EXAMINATION CARDS IN ARITHMETIC. (Dedicated to Lord Sandon.) With Answers and Hints.
Standards I. and II. in box, 1s. Standards III., IV., and V., in boxes, 1s. each. Standard VI. in Two Parts, in boxes, 1s. each.

A and B papers, of nearly the same difficulty, are given so as to prevent copying, and the colours of the A and B papers differ in each Standard, and from those of every other Standard, so that a master or mistress can see at a glance whether the children have the proper papers.

ALGEBRA.

Dalton.—RULES AND EXAMPLES IN ALGEBRA. By the Rev. T. DALTON, M.A., Assistant-Master of Eton College. Part I. New Edition. 18mo. 2s. Part II. 18mo. 2s. 6d.

Jones and Cheyne.—ALGEBRAICAL EXERCISES. Progressively Arranged. By the Rev. C. A. JONES, M.A., and C. H. CHEYNE, M.A., F.R.A.S., Mathematical Masters of Westminster School. New Edition. 18mo. 2s. 6d.

Smith.—ARITHMETIC AND ALGEBRA, in their Principles and Application; with numerous systematically arranged Examples taken from the Cambridge Examination Papers, with especial reference to the Ordinary Examination for the B.A. Degree. By the Rev. BARNARD SMITH, M.A., late Rector of Glaston, Rutland, and Fellow and Senior Bursar of St. Peter's College, Cambridge. New Edition, carefully Revised. Crown 8vo. 10s. 6d.

Todhunter.—Works by I. TODHUNTER, M.A., F.R.S., D.Sc., late of St. John's College, Cambridge.

"Mr. Todhunter is chiefly known to Students of Mathematics as the author of a series of admirable mathematical text-books, which possess the rare qualities of being clear in style and absolutely free from mistakes, typographical or other."—SATURDAY REVIEW.

ALGEBRA FOR BEGINNERS. With numerous Examples. New Edition. 18mo. 2s. 6d.
KEY TO ALGEBRA FOR BEGINNERS. Crown 8vo. 6s. 6d.
ALGEBRA. For the Use of Colleges and Schools. New Edition. Crown 8vo. 7s. 6d.
KEY TO ALGEBRA FOR THE USE OF COLLEGES AND SCHOOLS. Crown 8vo. 10s. 6d.

EUCLID & ELEMENTARY GEOMETRY.

Constable.—GEOMETRICAL EXERCISES FOR BEGINNERS. By SAMUEL CONSTABLE. Crown 8vo. 3s. 6d.

Cuthbertson.—EUCLIDIAN GEOMETRY. By FRANCIS CUTHBERTSON, M.A., LL.D., Head Mathematical Master of the City of London School. Extra fcap. 8vo. 4s. 6d.

MATHEMATICS.

Dodgson.—EUCLID. BOOKS I. AND II. Edited by CHARLES L. DODGSON, M.A., Student and late Mathematical Lecturer of Christ Church, Oxford. Second Edition, with words substituted for the Algebraical Symbols used in the First Edition. Crown 8vo. 2s.

*** The text of this Edition has been ascertained, by counting the words, to be *less than five-sevenths* of that contained in the ordinary editions.

Kitchener.—A GEOMETRICAL NOTE-BOOK, containing Easy Problems in Geometrical Drawing preparatory to the Study of Geometry. For the Use of Schools. By F. E. KITCHENER, M.A., Mathematical Master at Rugby. New Edition. 4to. 2s.

Mault.—NATURAL GEOMETRY: an Introduction to the Logical Study of Mathematics. For Schools and Technical Classes. With Explanatory Models, based upon the Tachymetrical works of Ed. Lagout. By A. MAULT. 18mo. 1s. Models to Illustrate the above, in Box, 12s. 6d.

Smith.— AN ELEMENTARY TREATISE ON SOLID GEOMETRY. By CHARLES SMITH, M.A., Fellow and Tutor of Sidney Sussex College, Cambridge. Crown 8vo. 9s. 6d.

Syllabus of Plane Geometry (corresponding to Euclid, Books I.—VI.). Prepared by the Association for the Improvement of Geometrical Teaching. New Edition. Crown 8vo. 1s.

Todhunter.—THE ELEMENTS OF EUCLID. For the Use of Colleges and Schools. By I. TODHUNTER, M.A., F.R.S., D.Sc., of St. John's College, Cambridge. New Edition. 18mo. 3s. 6d.
KEY TO EXERCISES IN EUCLID. Crown 8vo. 6s. 6d.

Wilson (J. M.).—ELEMENTARY GEOMETRY. BOOKS I.—V. Containing the Subjects of Euclid's first Six Books. Following the Syllabus of the Geometrical Association. By the Rev. J. M. WILSON, M.A., Head Master of Clifton College. New Edition. Extra fcap. 8vo. 4s. 6d.

MENSURATION.

Tebay.—ELEMENTARY MENSURATION FOR SCHOOLS. With numerous examples. By SEPTIMUS TEBAY, B.A., Head Master of Queen Elizabeth's Grammar School, Rivington. Extra fcap. 8vo. 3s. 6d.

Todhunter.—MENSURATION FOR BEGINNERS. By I. TODHUNTER, M.A., F.R.S., D.Sc., late of St. John's College, Cambridge. With Examples. New Edition. 18mo. 2s. 6d.

HIGHER MATHEMATICS.

Airy.—Works by Sir G. B. AIRY, K.C.B., formerly Astronomer-Royal:—

ELEMENTARY TREATISE ON PARTIAL DIFFERENTIAL EQUATIONS. Designed for the Use of Students in the Universities. With Diagrams. Second Edition. Crown 8vo. 5s. 6d.

ON THE ALGEBRAICAL AND NUMERICAL THEORY OF ERRORS OF OBSERVATIONS AND THE COMBINATION OF OBSERVATIONS. Second Edition, revised. Crown 8vo. 6s. 6d.

Alexander (T.).—ELEMENTARY APPLIED MECHANICS. Being the simpler and more practical Cases of Stress and Strain wrought out individually from first principles by means of Elementary Mathematics. By T. ALEXANDER, C.E., Professor of Civil Engineering in the Imperial College of Engineering, Tokei, Japan. Crown 8vo. Part I. 4s. 6d.

Alexander and Thomson.—ELEMENTARY APPLIED MECHANICS. By THOMAS ALEXANDER, C.E., Professor of Engineering in the Imperial College of Engineering, Tokei, Japan; and ARTHUR WATSON THOMSON, C.E., B.SC., Professor of Engineering at the Royal College, Cirencester. Part II. TRANSVERSE STRESS. Crown 8vo. 10s. 6d.

Bayma.—THE ELEMENTS OF MOLECULAR MECHANICS. By JOSEPH BAYMA, S.J., Professor of Philosophy, Stonyhurst College. Demy 8vo. 10s. 6d.

Beasley.—AN ELEMENTARY TREATISE ON PLANE TRIGONOMETRY. With Examples. By R. D. BEASLEY, M.A. Eighth Edition, revised and enlarged. Crown 8vo. 3s. 6d.

Blackburn (Hugh).—ELEMENTS OF PLANE TRIGONOMETRY, for the use of the Junior Class in Mathematics in the University of Glasgow. By HUGH BLACKBURN, M.A., late Professor of Mathematics in the University of Glasgow. Globe 8vo. 1s. 6d.

Boole.—Works by G. BOOLE, D.C.L., F.R.S., late Professor of Mathematics in the Queen's University, Ireland.

A TREATISE ON DIFFERENTIAL EQUATIONS. Supplementary Volume. Edited by I. TODHUNTER. Crown 8vo. 8s. 6d.

THE CALCULUS OF FINITE DIFFERENCES. Third Edition, revised by J. F. MOULTON. Crown 8vo. 10s. 6d.

Cambridge Senate-House Problems and Riders, with Solutions:—
  1875—PROBLEMS AND RIDERS. By A. G. GREENHILL, M.A. Crown 8vo. 8s. 6d.
  1878—SOLUTIONS OF SENATE-HOUSE PROBLEMS. By the Mathematical Moderators and Examiners. Edited by J. W. L. GLAISHER, M.A., Fellow of Trinity College, Cambridge. 12s.

Cheyne.—AN ELEMENTARY TREATISE ON THE PLANETARY THEORY. By C. H. H. CHEYNE, M.A., F.R.A.S. With a Collection of Problems. Third Edition. Edited by Rev. A. FREEMAN, M.A., F.R.A.S. Crown 8vo. 7s. 6d.

Christie.—A COLLECTION OF ELEMENTARY TEST-QUESTIONS IN PURE AND MIXED MATHEMATICS; with Answers and Appendices on Synthetic Division, and on the Solution of Numerical Equations by Horner's Method. By JAMES R. CHRISTIE, F.R.S., Royal Military Academy, Woolwich. Crown 8vo. 8s. 6d.

Clausius.—MECHANICAL THEORY OF HEAT. By R. CLAUSIUS. Translated by WALTER R. BROWNE, M.A., late Fellow of Trinity College, Cambridge. Crown 8vo. 10s. 6d.

Clifford.—THE ELEMENTS OF DYNAMIC. An Introduction to the Study of Motion and Rest in Solid and Fluid Bodies. By W. K. CLIFFORD, F.R.S., late Professor of Applied Mathematics and Mechanics at University College, London. Part I.—KINEMATIC. Crown 8vo. 7s. 6d.

Cotterill.—APPLIED MECHANICS: an Elementary General Introduction to the Theory of Structures and Machines. By JAMES H. COTTERILL, F.R.S., Associate Member of the Council of the Institution of Naval Architects, Associate Member of the Institution of Civil Engineers, Professor of Applied Mechanics in the Royal Naval College, Greenwich. Medium 8vo. 18s.

Day.—PROPERTIES OF CONIC SECTIONS PROVED GEOMETRICALLY. Part I. THE ELLIPSE. With Problems. By the Rev. H. G. DAY, M.A. 8vo. 3s. 6d.

Day (R. E.)—ELECTRIC LIGHT ARITHMETIC. By R. E. DAY, M.A., Evening Lecturer in Experimental Physics at King's College, London. Pott 8vo. 2s.

Drew.—GEOMETRICAL TREATISE ON CONIC SECTIONS. By W. H. DREW, M.A., St. John's College, Cambridge. New Edition, enlarged. Crown 8vo. 5s.
  SOLUTIONS TO THE PROBLEMS IN DREW'S CONIC SECTIONS. Crown 8vo. 4s. 6d.

Dyer.—EXERCISES IN ANALYTICAL GEOMETRY. Compiled and arranged by J. M. DYER, M.A., Senior Mathematical Master in the Classical Department of Cheltenham College. With Illustrations. Crown 8vo. 4s. 6d.

Eagles (T. H.).—A CONSTRUCTIVE TREATISE ON PLANE CURVES. By T. H. EAGLES, of the Royal Indian Engineering College, Cooper's Hill. With Illustrations. Crown 8vo.
[*In the press.*

Edgar (J. H.) and Pritchard (G. S.).—NOTE-BOOK ON PRACTICAL SOLID OR DESCRIPTIVE GEOMETRY. Containing Problems with help for Solutions. By J. H. EDGAR, M.A., Lecturer on Mechanical Drawing at the Royal School of Mines, and G. S. PRITCHARD. Fourth Edition, revised by ARTHUR MEEZE. Globe 8vo. 4s. 6d.

Ferrers.—Works by the Rev. N. M. FERRERS, M.A., Fellow and Master of Gonville and Caius College, Cambridge.
AN ELEMENTARY TREATISE ON TRILINEAR CO-ORDINATES, the Method of Reciprocal Polars, and the Theory of Projectors. New Edition, revised. Crown 8vo. 6s. 6d.
AN ELEMENTARY TREATISE ON SPHERICAL HARMONICS, AND SUBJECTS CONNECTED WITH THEM. Crown 8vo. 7s. 6d.

Forsyth.—A TREATISE ON DIFFERENTIAL EQUATIONS. By A. R. FORSYTH, M.A., Fellow of Trinity College, Cambridge. [*In preparation.*

Frost.—Works by PERCIVAL FROST, M.A., D.Sc., formerly Fellow of St. John's College, Cambridge; Mathematical Lecturer at King's College.
AN ELEMENTARY TREATISE ON CURVE TRACING. By PERCIVAL FROST, M.A. 8vo. 12s.
SOLID GEOMETRY. A New Edition, revised and enlarged, of the Treatise by FROST and WOLSTENHOLME. In 2 Vols. Vol. I. 8vo. 16s.

Hemming.—AN ELEMENTARY TREATISE ON THE DIFFERENTIAL AND INTEGRAL CALCULUS, for the Use of Colleges and Schools. By G. W. HEMMING, M.A., Fellow of St. John's College, Cambridge. Second Edition, with Corrections and Additions. 8vo. 9s.

Ibbetson.—A TREATISE ON ELASTICITY. By W. J. IBBETSON, M.A. Crown 8vo. [*In preparation.*

MATHEMATICS. 29

Jackson.—GEOMETRICAL CONIC SECTIONS. An Elementary Treatise in which the Conic Sections are defined as the Plane Sections of a Cone, and treated by the Method of Projection. By J. STUART JACKSON, M.A., late Fellow of Gonville and Caius College, Cambridge. Crown 8vo. 4s. 6d.

Jellet (John H.).—A TREATISE ON THE THEORY OF FRICTION. By JOHN H. JELLET, B.D., Provost of Trinity College, Dublin; President of the Royal Irish Academy. 8vo. 8s. 6d.

Johnson.—INTEGRAL CALCULUS, an Elementary Treatise on 'the; Founded on the Method of Rates or Fluxions. By WILLIAM WOOLSEY JOHNSON, Professor of Mathematics at the United States Naval Academy, Annopolis, Maryland. Demy 8vo. 8s.

Kelland and Tait.—INTRODUCTION TO QUATERNIONS, with numerous examples. By P. KELLAND, M.A., F.R.S., and P. G. TAIT, M.A., Professors in the Department of Mathematics in the University of Edinburgh. Second Edition. Crown 8vo. 7s. 6d.

Kempe.—HOW TO DRAW A STRAIGHT LINE: a Lecture on Linkages. By A. B. KEMPE. With Illustrations. Crown 8vo. 1s. 6d. (*Nature Series*.)

Knox—DIFFERENTIAL CALCULUS FOR BEGINNERS. By ALEXANDER KNOX. Fcap. 8vo. [*In the press.*

Lock.—ELEMENTARY TRIGONOMETRY. By Rev. J. B. LOCK, M.A., Senior Fellow, Assistant Tutor and Lecturer in Mathematics, of Gonville and Caius College, Cambridge; late Assistant-Master at Eton. Globe 8vo. 4s. 6d.
HIGHER TRIGONOMETRY. By the same Author. Globe 8vo. 3s. 6d.
Both Parts complete in One Volume. Globe 8vo. 7s. 6d.

Lupton.—ELEMENTARY CHEMICAL ARITHMETIC. With 1,100 Problems. By SYDNEY LUPTON, M.A., Assistant-Master in Harrow School. Globe 8vo. 5s.

Macfarlane.—PHYSICAL ARITHMETIC. By ALEXANDER MACFARLANE, D.Sc., Examiner in Mathematics in the University of Edinburgh. Crown 8vo. [*In the press.*

Merriman.—ELEMENTS OF THE METHOD OF LEAST SQUARE. By MANSFIELD MERRIMAN, Ph.D., Professor of Civil and Mechanical Engineering, Lehigh University, Bethlehem, Penn. Crown 8vo. 7s. 6d.

Millar.—ELEMENTS OF DESCRIPTIVE GEOMETRY. By J. B. MILLAR, C.E., Assistant Lecturer in Engineering in Owens College, Manchester. Crown 8vo. 6s.

Milne.—WEEKLY PROBLEM PAPERS. By the Rev. JOHN J. MILNE, M.A., Second Master of Heversham Grammar School, Member of the London Mathematical Society, Member of the Association for the Improvement of Geometrical Teaching, late Scholar of St. John's College, Cambridge. Pott 8vo.
[In the press.

Morgan.—A COLLECTION OF PROBLEMS AND EXAMPLES IN MATHEMATICS. With Answers. By H. A. MORGAN, M.A., Sadlerian and Mathematical Lecturer of Jesus College, Cambridge. Crown 8vo. 6s. 6d.

Muir.—A TREATISE ON THE THEORY OF DETERMINANTS. With graduated sets of Examples. For use in Colleges and Schools. By THOS. MUIR, M.A., F.R.S.E., Mathematical Master in the High School of Glasgow. Crown 8vo. 7s. 6d.

Parkinson.—AN ELEMENTARY TREATISE ON MECHANICS. For the Use of the Junior Classes at the University and the Higher Classes in Schools. By S. PARKINSON, D.D., F.R.S., Tutor and Prælector of St. John's College, Cambridge. With a Collection of Examples. Sixth Edition, revised. Crown 8vo. 9s. 6d.

Phear.—ELEMENTARY HYDROSTATICS. With Numerous Examples. By J. D. PHEAR, M.A., Fellow and late Assistant Tutor of Clare College, Cambridge. New Edition. Crown 8vo. 5s. 6d.

Pirie.—LESSONS ON RIGID DYNAMICS. By the Rev. G. PIRIE, M.A., late Fellow and Tutor of Queen's College, Cambridge; Professor of Mathematics in the University of Aberdeen. Crown 8vo. 6s.

Puckle.—AN ELEMENTARY TREATISE ON CONIC SECTIONS AND ALGEBRAIC GEOMETRY. With Numerou Examples and Hints for their Solution; especially designed for the Use of Beginners. By G. H. PUCKLE, M.A. Fifth Edition, revised and enlarged. Crown 8vo. 7s. 6d.

Rawlinson.—ELEMENTARY STATICS. By the Rev. GEORGE RAWLINSON, M.A. Edited by the Rev. EDWARD STURGES, M.A. Crown 8vo. 4s. 6d.

Reynolds.—MODERN METHODS IN ELEMENTARY GEOMETRY. By E. M. REYNOLDS, M.A., Mathematical Master in Clifton College. Crown 8vo. 3s. 6d.

Reuleaux.—THE KINEMATICS OF MACHINERY. Outlines of a Theory of Machines. By Professor F. REULEAUX. Translated and Edited by Professor A. B. W. KENNEDY, C.E. With 450 Illustrations. Medium 8vo. 21s.

Rice and Johnson.—DIFFERENTIAL CALCULUS, an Elementary Treatise on the ; Founded on the Method of Rates or Fluxions. By JOHN MINOT RICE, Professor of Mathematics in the United States Navy, and WILLIAM WOOLSEY JOHNSON, Professor of Mathematics at the United States Naval Academy. Third Edition, Revised and Corrected. Demy 8vo. 16s. Abridged Edition, 8s.

Robinson.—TREATISE ON MARINE SURVEYING. Prepared for the use of younger Naval Officers. With Questions for Examinations and Exercises principally from the Papers of the Royal Naval College. With the results. By Rev. JOHN L. ROBINSON, Chaplain and Instructor in the Royal Naval College, Greenwich. With Illustrations. Crown 8vo. 7s. 6d.
CONTENTS.—Symbols used in Charts and Surveying—The Construction and Use of Scales—Laying off Angles—Fixing Positions by Angles — Charts and Chart-Drawing—Instruments and Observing — Base Lines—Triangulation—Levelling—Tides and Tidal Observations—Soundings—Chronometers—Meridian Distances —Method of Plotting a Survey—Miscellaneous Exercises—Index.

Routh.—Works by EDWARD JOHN ROUTH, M.A., F.R.S., D.Sc., late Fellow and Assistant Tutor at St. Peter's College, Cambridge; Examiner in the University of London.
A TREATISE ON THE DYNAMICS OF THE SYSTEM OF RIGID BODIES. With numerous Examples. Fourth and enlarged Edition. Two Vols. Vol. I.—Elementary Parts. 8vo. 14s. Vol. II.—The Advance Parts. 8vo. *[Just ready.*
STABILITY OF A GIVEN STATE OF MOTION, PARTICULARLY STEADY MOTION. Adams' Prize Essay for 1877. 8vo. 8s. 6d.

Smith (C.).—Works by CHARLES SMITH, M.A., Fellow and Tutor of Sidney Sussex College, Cambridge.
CONIC SECTIONS. Second Edition. Crown 8vo. 7s. 6d.
AN ELEMENTARY TREATISE ON SOLID GEOMETRY. Crown 8vo. 9s. 6d.

Snowball.—THE ELEMENTS OF PLANE AND SPHERICAL TRIGONOMETRY; with the Construction and Use of Tables of Logarithms. By J. C. SNOWBALL, M.A. New Edition. Crown 8vo. 7s. 6d.

Tait and Steele.—A TREATISE ON DYNAMICS OF A PARTICLE. With numerous Examples. By Professor TAIT and Mr. STEELE. Fourth Edition, revised. Crown 8vo. 12s.

Thomson.—A TREATISE ON THE MOTION OF VORTEX RINGS. An Essay to which the Adams Prize was adjudged in 1882 in the University of Cambridge. By J. J. THOMSON, Fellow and Assistant Lecturer of Trinity College, Cambridge. With Diagrams. 8vo. 6s.

Todhunter.—Works by I. TODHUNTER, M.A., F.R.S., D.Sc., late of St. John's College, Cambridge.

"Mr. Todhunter is chiefly known to students of Mathematics as the author of a series of admirable mathematical text-books, which possess the rare qualities of being clear in style and absolutely free from mistakes, typographical and other."— SATURDAY REVIEW.

TRIGONOMETRY FOR BEGINNERS. With numerous Examples. New Edition. 18mo. 2s. 6d.

KEY TO TRIGONOMETRY FOR BEGINNERS. Crown 8vo. 8s. 6d.

MECHANICS FOR BEGINNERS. With numerous Examples. New Edition. 18mo. 4s. 6d.

KEY TO MECHANICS FOR BEGINNERS. Crown 8vo. 6s. 6d.

AN ELEMENTARY TREATISE ON THE THEORY OF EQUATIONS. New Edition, revised. Crown 8vo. 7s. 6d.

PLANE TRIGONOMETRY. For Schools and Colleges. New Edition. Crown 8vo. 5s.

KEY TO PLANE TRIGONOMETRY. Crown 8vo. 10s. 6d.

A TREATISE ON SPHERICAL TRIGONOMETRY. New Edition, enlarged. Crown 8vo. 4s. 6d.

PLANE CO-ORDINATE GEOMETRY, as applied to the Straight Line and the Conic Sections. With numerous Examples. New Edition, revised and enlarged. Crown 8vo. 7s. 6d.

A TREATISE ON THE DIFFERENTIAL CALCULUS. With numerous Examples. New Edition. Crown 8vo. 10s. 6d.

A TREATISE ON THE INTEGRAL CALCULUS AND ITS APPLICATIONS. With numerous Examples. New Edition, revised and enlarged. Crown 8vo. 10s. 6d.

EXAMPLES OF ANALYTICAL GEOMETRY OF THREE DIMENSIONS. New Edition, revised. Crown 8vo. 4s.

A TREATISE ON ANALYTICAL STATICS. With numerous Examples. New Edition, revised and enlarged. Crown 8vo. 10s. 6d.

Todhunter.—Works by I. TODHUNTER, M.A., &c. (*continued*)—
A HISTORY OF THE MATHEMATICAL THEORY OF PROBABILITY, from the time of Pascal to that of Laplace. 8vo. 18s.
RESEARCHES IN THE CALCULUS OF VARIATIONS, principally on the Theory of Discontinuous Solutions: an Essay to which the Adams' Prize was awarded in the University of Cambridge in 1871. 8vo. 6s.
A HISTORY OF THE MATHEMATICAL THEORIES OF ATTRACTION, AND THE FIGURE OF THE EARTH, from the time of Newton to that of Laplace. 2 vols. 8vo. 24s.
AN ELEMENTARY TREATISE ON LAPLACE'S, LAME'S, AND BESSEL'S FUNCTIONS. Crown 8vo. 10s. 6d.

Wilson (J. M.).—SOLID GEOMETRY AND CONIC SECTIONS. With Appendices on Transversals and Harmonic Division. For the Use of Schools. By Rev. J. M. WILSON, M.A. Head Master of Clifton College. New Edition. Extra fcap. 8vo. 3s. 6d.

Wilson.—GRADUATED EXERCISES IN PLANE TRIGONOMETRY. Compiled and arranged by J. WILSON, M.A., and S. R. WILSON, B.A. Crown 8vo. 4s. 6d.
"The exercises seem beautifully graduated and adapted to lead a student on most gently and pleasantly."—E. J. ROUTH, F.R.S., St. Peter's College, Cambridge.
(See also *Elementary Geometry*.)

Wilson (W. P.).—A TREATISE ON DYNAMICS. By W. P. WILSON, M.A., Fellow of St. John's College, Cambridge, and Professor of Mathematics in Queen's College, Belfast. 8vo. 9s. 6d.

Woolwich Mathematical Papers, for Admission into the Royal Military Academy, Woolwich, 1880—1883 inclusive. Crown 8vo. 3s. 6d.

Wolstenholme.—MATHEMATICAL PROBLEMS, on Subjects included in the First and Second Divisions of the Schedule of subjects for the Cambridge Mathematical Tripos Examination. Devised and arranged by JOSEPH WOLSTENHOLME, D.Sc., late Fellow of Christ's College, sometime Fellow of St. John's College, and Professor of Mathematics in the Royal Indian Engineering College. New Edition, greatly enlarged. 8vo. 18s.
EXAMPLES FOR PRACTICE IN THE USE OF SEVEN FIGURE LOGARITHMS. By the same Author. [*In preparation*.

SCIENCE.

(1) Natural Philosophy, (2) Astronomy, (3) Chemistry, (4) Biology, (5) Medicine, (6) Anthropology, (7) Physical Geography and Geology, (8) Agriculture, (9) Political Economy, (10) Mental and Moral Philosophy.

NATURAL PHILOSOPHY.

Airy.—Works by Sir G. B. AIRY, K.C.B., formerly Astronomer-Royal :—

UNDULATORY THEORY OF OPTICS. Designed for the Use of Students in the University. New Edition. Crown 8vo. 6s. 6d.

ON SOUND AND ATMOSPHERIC VIBRATIONS. With the Mathematical Elements of Music. Designed for the Use of Students in the University. Second Edition, revised and enlarged. Crown 8vo. 9s.

A TREATISE ON MAGNETISM. Designed for the Use of Students in the University. Crown 8vo. 9s. 6d.

GRAVITATION: an Elementary Explanation of the Principal Perturbations in the Solar System. New Edition. Crown 8vo.
[*Just ready.*

Airy (Osmond).—A TREATISE ON GEOMETRICAL OPTICS. Adapted for the Use of the Higher Classes in Schools. By OSMUND AIRY, B.A., one of the Mathematical Masters in Wellington College. Extra fcap. 8vo. 3s. 6d.

Alexander (T.).—ELEMENTARY APPLIED MECHANICS. Being the simpler and more practical Cases of Stress and Strain wrought out individually from first principles by means of Elementary Mathematics. By T. ALEXANDER, C.E., Professor of Civil Engineering in the Imperial College of Engineering, Tokei, Japan. Crown 8vo. Part I. 4s. 6d.

Alexander — Thomson. — ELEMENTARY APPLIED MECHANICS. By THOMAS ALEXANDER, C.E., Professor of Engineering in the Imperial College of Engineering, Tokei, Japan; and ARTHUR WATSON THOMSON, C.E., B.Sc., Professor of Engineering at the Royal College, Cirencester. Part II. TRANSVERSE STRESS; upwards of 150 Diagrams, and 200 Examples carefully worked out; new and complete method for finding, at every point of a beam, the amount of the greatest bending moment and shearing force during the transit of any set of loads fixed relatively to one another—*e.g.*, the wheels of a locomotive; continuous beams, &c., &c. Crown 8vo. 10s. 6d.

SCIENCE.

Awdry.—EASY LESSONS ON LIGHT. By Mrs. W. AWDRY. Illustrated. Extra fcap. 8vo. 2s. 6d.

Ball (R. S.).—EXPERIMENTAL MECHANICS. A Course of Lectures delivered at the Royal College of Science for Ireland. By R. S. BALL, M.A., Professor of Applied Mathematics and Mechanics in the Royal College of Science for Ireland. Cheaper Issue. Royal 8vo. 10s. 6d.

Chisholm.—THE SCIENCE OF WEIGHING AND MEASURING, AND THE STANDARDS OF MEASURE AND WEIGHT. By H.W. CHISHOLM, Warden of the Standards. With numerous Illustrations. Crown 8vo. 4s. 6d. (*Nature Series.*)

Clausius.—MECHANICAL THEORY OF HEAT. By R. CLAUSIUS. Translated by WALTER R. BROWNE, M.A., late Fellow of Trinity College, Cambridge. Crown 8vo. 10s. 6d.

Cotterill.—APPLIED MECHANICS: an Elementary General Introduction to the Theory of Structures and Machines. By JAMES H. COTTERILL, F.R.S., Associate Member of the Council of the Institution of Naval Architects, Associate Member of the Institution of Civil Engineers, Professor of Applied Mechanics in the Royal Naval College, Greenwich. Medium 8vo. 18s.

Cumming.—AN INTRODUCTION TO THE THEORY OF ELECTRICITY: By LINNÆUS CUMMING, M.A., one of the Masters of Rugby School. With Illustrations. Crown 8vo. 8s. 6d.

Daniell.—A TEXT-BOOK OF THE PRINCIPLES OF PHYSICS. By ALFRED DANIELL, M.A., D.Sc., Lecturer on Physics in the School of Medicine, Edinburgh. With Illustrations. Medium 8vo. 21s.

Day.—ELECTRIC LIGHT ARITHMETIC. By R. E. DAY, M.A., Evening Lecturer in Experimental Physics at King's College, London. Pott 8vo. 2s.

Everett.—UNITS AND PHYSICAL CONSTANTS. By J. D. EVERETT, F.R.S., Professor of Natural Philosophy, Queen's College, Belfast. Extra fcap. 8vo. 4s. 6d.

Gray.—ABSOLUTE MEASUREMENTS IN ELECTRICITY AND MAGNETISM. By ANDREW GRAY, M.A., F.R.S.E., Professor of Physics in the University College of North Wales. Pott 8vo. 3s. 6d.

Grove.—A DICTIONARY OF MUSIC AND MUSICIANS. By Eminent Writers, English and Foreign. Edited by Sir GEORGE GROVE, D.C.L., Director of the Royal College of Music, &c. Demy 8vo.
 Vols. I., II., and III. Price 21s. each.
 Vol. I. A to IMPROMPTU. Vol. II. IMPROPERIA to PLAIN SONG. Vol. III. PLANCHE TO SUMER IS ICUMEN IN. Demy 8vo. cloth, with Illustrations in Music Type and Woodcut. Also published in Parts. Parts I. to XIV., and Part XIX., price 3s. 6d. each. Parts XV., XVI., price 7s. Parts XVII., XVIII., price 7s.

"Dr. Grove's Dictionary will be a boon to every intelligent lover of music."—*Saturday Review.*

Huxley.—INTRODUCTORY PRIMER OF SCIENCE. By T. H. HUXLEY, P.R.S., Professor of Natural History in the Royal School of Mines, &c. 18mo. 1s.

Kempe.—HOW TO DRAW A STRAIGHT LINE; a Lecture on Linkages. By A. B. KEMPE. With Illustrations. Crown 8vo. 1s. 6d. (*Nature Series.*)

Kennedy.—MECHANICS OF MACHINERY. By A. B. W. KENNEDY, M.Inst.C.E., Professor of Engineering and Mechanical Technology in University College, London. With Illustrations. Crown 8vo. [*In the press.*

Lang.—EXPERIMENTAL PHYSICS. By P. R. SCOTT LANG. M.A., Professor of Mathematics in the University of St. Andrews. Crown 8vo. [*In preparation.*

Lupton.—NUMERICAL TABLES AND CONSTANTS IN ELEMENTARY SCIENCE. By SYDNEY LUPTON, M.A., F.C.S., F.I.C., Assistant Master at Harrow School. Extra fcap. 8vo. 2s. 6d.

Macfarlane.— HYSICAL ARITHMETIC. By ALEXANDER MACFARLANE, D.Sc., Examiner in Mathematics in the University of Edinburgh. [*In the press.*

Martineau (Miss C. A.).—EASY LESSONS ON HEAT. By Miss C. A. MARTINEAU. Illustrated. Extra fcap. 8vo. 2s. 6d.

Mayer.—SOUND : a Series of Simple, Entertaining, and Inexpensive Experiments in the Phenomena of Sound, for the Use of Students of every age. By A. M. MAYER, Professor of Physics in the Stevens Institute of Technology, &c. With numerous Illustrations. Crown 8vo. 2s. 6d. (*Nature Series.*)

SCIENCE.

Mayer and Barnard.—LIGHT: a Series of Simple, Entertaining, and Inexpensive Experiments in the Phenomena of Light, for the Use of Students of every age. By A. M. MAYER and C. BARNARD. With numerous Illustrations. Crown 8vo. 2s. 6d. (*Nature Series.*)

Newton.—PRINCIPIA. Edited by Professor Sir W. THOMSON and Professor BLACKBURNE. 4to, cloth. 31s. 6d.

THE FIRST THREE SECTIONS OF NEWTON'S PRINCIPIA. With Notes and Illustrations. Also a Collection of Problems, principally intended as Examples of Newton's Methods. By PERCIVAL FROST, M.A. Third Edition. 8vo. 12s.

Parkinson.—A TREATISE ON OPTICS. By S. PARKINSON, D.D., F.R.S., Tutor and Prælector of St. John's College, Cambridge. Fourth Edition, revised and enlarged. Crown 8vo. 10s. 6d.

Perry.—STEAM. AN ELEMENTARY TREATISE. By JOHN PERRY, C.E., Whitworth Scholar, Fellow of the Chemical Society, Lecturer in Physics at Clifton College. With numerous Woodcuts and Numerical Examples and Exercises. 18mo. 4s. 6d.

Ramsay.—EXPERIMENTAL PROOFS OF CHEMICAL THEORY FOR BEGINNERS. By WILLIAM RAMSAY, Ph.D., Professor of Chemistry in University College, Bristol. Pott 8vo. 2s. 6d.

Rayleigh.—THE THEORY OF SOUND. By LORD RAYLEIGH, M.A., F.R.S., formerly Fellow of Trinity College, Cambridge, 8vo. Vol. I. 12s. 6d. Vol. II. 12s. 6d. [Vol. III. *in the press.*

Reuleaux.—THE KINEMATICS OF MACHINERY. Outlines of a Theory of Machines. By Professor F. REULEAUX. Translated and Edited by Professor A. B. W. KENNEDY, C.E. With 450 Illustrations. Medium 8vo. 21s.

Shann.—AN ELEMENTARY TREATISE ON HEAT, IN RELATION TO STEAM AND THE STEAM-ENGINE. By G. SHANN, M.A. With Illustrations. Crown 8vo. 4s. 6d.

Spottiswoode.—POLARISATION OF LIGHT. By the late W. SPOTTISWOODE, P.R.S. With many Illustrations. New Edition. Crown 8vo. 3s. 6d. (*Nature Series.*)

Stewart (Balfour).—Works by BALFOUR STEWART, F.R.S., Professor of Natural Philosophy in the Victoria University the Owens College, Manchester.

PRIMER OF PHYSICS. With numerous Illustrations. New Edition, with Questions. 18mo. 1s. (*Science Primers.*)

Stewart (Balfour).—Works by *(continued)*—
LESSONS IN ELEMENTARY PHYSICS. With numerous Illustrations and Chromolitho of the Spectra of the Sun, Stars, and Nebulæ. New Edition. Fcap. 8vo. 4s. 6d.
QUESTIONS ON BALFOUR STEWART'S ELEMENTARY LESSONS IN PHYSICS. By Prof. THOMAS H. CORE, Owens College, Manchester. Fcap. 8vo. 2s.

Stewart—Gee.—PRACTICAL PHYSICS, ELEMENTARY LESSONS IN. By Professor BALFOUR STEWART, F.R.S., and W. HALDANE GEE. Fcap. 8vo.
Part I. General Physics. [*Nearly ready.*
Part II. Optics, Heat, and Sound. [*In preparation.*
Part III. Electricity and Magnetism. [*In preparation.*

Stokes.—ON LIGHT. Burnett Lectures. First Course. ON THE NATURE OF LIGHT. Delivered in Aberdeen in November 1883. By GEORGE GABRIEL STOKES, M.A., F.R.S., &c., Fellow of Pembroke College, and Lucasian Professor of Mathematics in the University of Cambridge. Crown 8vo. 2s. 6d.

Stone.—AN ELEMENTARY TREATISE ON SOUND. By W. H. STONE, M.B. With Illustrations. 18mo. 3s. 6d.

Tait.—HEAT. By P. G. TAIT, M.A., Sec. R.S.E., Formerly Fellow of St. Peter's College, Cambridge, Professor of Natural Philosophy in the University of Edinburgh. Crown 8vo. 6s.

Thompson.—ELEMENTARY LESSONS IN ELECTRICITY AND MAGNETISM. By SILVANUS P. THOMPSON. Professor of Experimental Physics in University College, Bristol. With Illustrations. Fcap. 8vo. 4s. 6d.

Thomson.—ELECTROSTATICS AND MAGNETISM, REPRINTS OF PAPERS ON. By Sir WILLIAM THOMSON, D.C.L., LL.D., F.R.S., F.R.S.E., Fellow of St. Peter's College, Cambridge, and Professor of Natural Philosophy in the University of Glasgow. Second Edition. Medium 8vo. 18s.

Thomson.—THE MOTION OF VORTEX RINGS, A TREATISE ON. An Essay to which the Adams Prize was adjudged in 1882 in the University of Cambridge. By J. J. THOMSON, Fellow and Assistant-Lecturer of Trinity College, Cambridge. With Diagrams. 8vo. 6s.

Todhunter.—NATURAL PHILOSOPHY FOR BEGINNERS. By I. TODHUNTER, M.A., F.R.S., D.Sc.
Part I. The Properties of Solid and Fluid Bodies. 18mo. 3s. 6d.
Part II. Sound, Light, and Heat. 18mo. 3s. 6d.

Wright (Lewis).—LIGHT; A COURSE OF EXPERIMENTAL OPTICS, CHIEFLY WITH THE LANTERN. By LEWIS WRIGHT. With nearly 200 Engravings and Coloured Plates. Crown 8vo. 7s. 6d.

ASTRONOMY.

Airy.—POPULAR ASTRONOMY. With Illustrations by Sir G. B. AIRY, K.C.B., formerly Astronomer-Royal. New Edition. 18mo. 4s. 6d.

Forbes.—TRANSIT OF VENUS. By G. FORBES, M.A., Professor of Natural Philosophy in the Andersonian University, Glasgow. Illustrated. Crown 8vo. 3s. 6d. (*Nature Series.*)

Godfray.—Works by HUGH GODFRAY, M.A., Mathematical Lecturer at Pembroke College, Cambridge.
A TREATISE ON ASTRONOMY, for the Use of Colleges and Schools. New Edition. 8vo. 12s. 6d.
AN ELEMENTARY TREATISE ON THE LUNAR THEORY, with a Brief Sketch of the Problem up to the time of Newton. Second Edition, revised. Crown 8vo. 5s. 6d.

Lockyer.—Works by J. NORMAN LOCKYER, F.R.S.
PRIMER OF ASTRONOMY. With numerous Illustrations. New Edition. 18mo. 1s. (*Science Primers.*)
ELEMENTARY LESSONS IN ASTRONOMY. With Coloured Diagram of the Spectra of the Sun, Stars, and Nebulæ, and numerous Illustrations. New Edition. Fcap. 8vo. 5s. 6d.
QUESTIONS ON LOCKYER'S ELEMENTARY LESSONS IN ASTRONOMY. For the Use of Schools. By JOHN FORBES-ROBERTSON. 18mo, cloth limp. 1s. 6d.
THE SPECTROSCOPE AND ITS APPLICATIONS. With Coloured Plate and numerous Illustrations. New Edition. Crown 8vo. 3s. 6d.

Newcomb.—POPULAR ASTRONOMY. By S. NEWCOMB, LL.D., Professor U.S. Naval Observatory. With 112 Illustrations and 5 Maps of the Stars. Second Edition, revised. 8vo. 18s.

"It is unlike anything else of its kind, and will be of more use in circulating a knowledge of Astronomy than nine-tenths of the books which have appeared on th subject of late years."—SATURDAY REVIEW.

CHEMISTRY.

Fleischer.—A SYSTEM OF VOLUMETRIC ANALYSIS. Translated, with Notes and Additions, from the Second German Edition, by M. M. PATTISON MUIR, F.R.S.E. With Illustrations. Crown 8vo. 7s. 6d.

Jones.—Works by Francis Jones, F.R.S.E., F.C.S., Chemical Master in the Grammar School, Manchester.

THE OWENS COLLEGE JUNIOR COURSE OF PRACTICAL CHEMISTRY. With Preface by Sir Henry Roscoe, and Illustrations. New Edition. 18mo. 2s. 6d.

QUESTIONS ON CHEMISTRY. A Series of Problems and Exercises in Inorganic and Organic Chemistry. Fcap. 8vo. 3s.

Landauer.—BLOWPIPE ANALYSIS. By J. Landauer. Authorised English Edition by J. Taylor and W. E. Kay, of Owens College, Manchester. Extra fcap. 8vo. 4s. 6d.

Lupton.—ELEMENTARY CHEMICAL ARITHMETIC. With 1,100 Problems. By Sydney Lupton, M.A., Assistant-Master at Harrow. Extra fcap. 8vo. 5s.

Muir.—PRACTICAL CHEMISTRY FOR MEDICAL STUDENTS. Specially arranged for the first M.B. Course. By M. M. Pattison Muir, F.R.S.E. Fcap. 8vo. 1s. 6d.

Roscoe.—Works by Sir Henry E. Roscoe, F.R.S., Professor of Chemistry in the Victoria University the Owens College, Manchester.

PRIMER OF CHEMISTRY. With numerous Illustrations. New Edition. With Questions. 18mo. 1s. (*Science Primers*).

LESSONS IN ELEMENTARY CHEMISTRY, INORGANIC AND ORGANIC. With numerous Illustrations and Chromolitho of the Solar Spectrum, and of the Alkalies and Alkaline Earths. New Edition. Fcap. 8vo. 4s. 6d.

A SERIES OF CHEMICAL PROBLEMS, prepared with Special Reference to the foregoing, by T. E. Thorpe, Ph.D., Professor of Chemistry in the Yorkshire College of Science, Leeds, Adapted for the Preparation of Students for the Government, Science, and Society of Arts Examinations. With a Preface by Sir Henry E. Roscoe, F.R.S. New Edition, with Key. 18mo. 2s.

Roscoe and Schorlemmer.—INORGANIC AND ORGANIC CHEMISTRY. A Complete Treatise on Inorganic and Organic Chemistry. By Sir Henry E. Roscoe, F.R.S., and Professor C. Schorlemmer, F.R.S. With numerous Illustrations. Medium 8vo.

Vols. I. and II.—INORGANIC CHEMISTRY.

Vol. I.—The Non-Metallic Elements. 21s. Vol. II. Part I.—Metals. 18s. Vol. II. Part II.—Metals. 18s.

Vol. III.—ORGANIC CHEMISTRY. Two Parts.

THE CHEMISTRY OF THE HYDROCARBONS and their Derivatives, or ORGANIC CHEMISTRY. With numerous Illustrations. Medium 8vo. 21s. each.

Schorlemmer.—A MANUAL OF THE CHEMISTRY OF THE CARBON COMPOUNDS, OR ORGANIC CHEMISTRY. By C. SCHORLEMMER, F.R.S., Professor of Chemistry in the Victoria University the Owens College, Manchester. With Illustrations. 8vo. 14s.

Thorpe.—A SERIES OF CHEMICAL PROBLEMS, prepared with Special Reference to Sir H. Roscoe's Lessons in Elementary Chemistry, by T. E. THORPE, Ph.D., Professor of Chemistry in the Yorkshire College of Science, Leeds, adapted for the Preparation of Students for the Government, Science, and Society of Arts Examinations. With a Preface by Sir HENRY E. ROSCOE. New Edition, with Key. 18mo. 2s.

Thorpe and Rücker.—A TREATISE ON CHEMICAL PHYSICS. By Professor THORPE, F.R.S., and Professor RÜCKER, of the Yorkshire College of Science. Illustrated. 8vo. [*In preparation.*]

Wright.—METALS AND THEIR CHIEF INDUSTRIAL APPLICATIONS. By C. ALDER WRIGHT, D.Sc., &c., Lecturer on Chemistry in St. Mary's Hospital Medical School. Extra fcap. 8vo. 3s. 6d.

BIOLOGY.

Allen.—ON THE COLOUR OF FLOWERS, as Illustrated in the British Flora. By GRANT ALLEN. With Illustrations. Crown 8vo. 3s. 6d. (*Nature Series.*)

Balfour.— A TREATISE ON COMPARATIVE EMBRYOLOGY. By F. M. BALFOUR, M.A., F.R.S., Fellow and Lecturer of Trinity College, Cambridge. With Illustrations. In 2 vols. 8vo. Vol. I. 18s. Vol. II. 21s.

Bettany.—FIRST LESSONS IN PRACTICAL BOTANY. By G. T. BETTANY, M.A., F.L.S., Lecturer in Botany at Guy's Hospital Medical School. 18mo. 1s.

Darwin (Charles).—MEMORIAL NOTICES OF CHARLES DARWIN, F.R.S., &c. By Professor HUXLEY, P.R.S., G. J. ROMANES, F.R.S., ARCHIBALD GEIKIE, F.R.S., and W. T. THISELTON DYER, F.R.S. Reprinted from *Nature*. With a Portrait, engraved by C. H. JEENS. Crown 8vo. 2s. 6d. (*Nature Series.*)

Dyer and Vines.—THE STRUCTURE OF PLANTS. By Professor THISELTON DYER, F.R.S., assisted by SYDNEY VINES, D.Sc., Fellow and Lecturer of Christ's College, Cambridge, and F. O. BOWER, M.A., Lecturer in the Normal School of Science. With numerous Illustrations. [*In the press.*

Flower (W. H.)—AN INTRODUCTION TO THE OSTEOLOGY OF THE MAMMALIA. Being the substance of the Course of Lectures delivered at the Royal College of Surgeons of England in 1870. By Professor W. H. FLOWER, F.R.S., F.R.C.S. With numerous Illustrations. New Edition, enlarged. Crown 8vo. 10s. 6d.

Foster.—Works by MICHAEL FOSTER, M.D., Sec. R.S., Professor of Physiology in the University of Cambridge.
PRIMER OF PHYSIOLOGY. With numerous Illustrations. New Edition. 18mo. 1s.
A TEXT-BOOK OF PHYSIOLOGY. With Illustrations. Fourth Edition, revised. 8vo. 21s.

Foster and Balfour.—THE ELEMENTS OF EMBRYOLOGY. By MICHAEL FOSTER, M.A., M.D., LL.D., Sec. R.S., Professor of Physiology in the University of Cambridge, Fellow of Trinity College, Cambridge, and the late FRANCIS M. BALFOUR, M.A., LL.D., F.R.S., Fellow of Trinity College, Cambridge, and Professor of Animal Morphology in the University. Second Edition, revised. Edited by ADAM SEDGWICK, M.A., Fellow and Assistant Lecturer of Trinity College, Cambridge, and WALTER HEAPE, Demonstrator in the Morphological Laboratory of the University of Cambridge. With Illustrations. Crown 8vo. 10s. 6d.

Foster and Langley.—A COURSE OF ELEMENTARY PRACTICAL PHYSIOLOGY. By Prof. MICHAEL FOSTER, M.D., Sec. R.S., &c., and J. N. LANGLEY, M.A., F.R.S., Fellow of Trinity College, Cambridge. Fifth Edition. Crown 8vo. 7s. 6d.

Gamgee.—A TEXT-BOOK OF THE PHYSIOLOGICAL CHEMISTRY OF THE ANIMAL BODY. Including an Account of the Chemical Changes occurring in Disease. By A. GAMGEE, M.D., F.R.S., Professor of Physiology in the Victoria University the Owens College, Manchester. 2 Vols. 8vo. With Illustrations. Vol. I. 18s. [*Vol. II. in the press.*

Gegenbaur.—ELEMENTS OF COMPARATIVE ANATOMY. By Professor CARL GEGENBAUR. A Translation by F. JEFFREY BELL, B.A. Revised with Preface by Professor E. RAY LANKESTER, F.R.S. With numerous Illustrations. 8vo. 21s.

Gray.—STRUCTURAL BOTANY, OR ORGANOGRAPHY ON THE BASIS OF MORPHOLOGY. To which are added the principles of Taxonomy, and Phytography, and a Glossary of Botanical Terms. By Professor ASA GRAY, LL.D. 8vo. 10s. 6d.

Hooker.—Works by Sir J. D. HOOKER, K.C.S.I., C.B., M.D., F.R.S., D.C.L.
PRIMER OF BOTANY. With numerous Illustrations. New Edition. 18mo. 1s. (*Science Primers.*)
THE STUDENT'S FLORA OF THE BRITISH ISLANDS- Third Edition, revised. Globe 8vo. 10s. 6d.

Howes.—AN ATLAS OF BIOLOGY. By E. B. HOWES, Demonstrator in the Science and Art Department, South Kensington. 4to. [*In the press.*

Huxley.—Works by Professor HUXLEY, P.R.S.
INTRODUCTORY PRIMER OF SCIENCE. 18mo. 1s. (*Science Primers.*)
LESSONS IN ELEMENTARY PHYSIOLOGY. With numerous Illustrations. New Edition. Fcap. 8vo. 4s. 6d.
QUESTIONS ON HUXLEY'S PHYSIOLOGY FOR SCHOOLS. By T. ALCOCK, M.D. 18mo. 1s. 6d.
PRIMER OF ZOOLOGY. 18mo. (*Science Primers.*)
[*In preparation.*

Huxley and Martin.—A COURSE OF PRACTICAL INSTRUCTION IN ELEMENTARY BIOLOGY. By Professor HUXLEY, P.R.S., assisted by H. N. MARTIN, M.B., D.Sc. New Edition, revised. Crown 8vo. 6s.

Lankester.—Works by Professor E. RAY LANKESTER, F.R.S.
A TEXT BOOK OF ZOOLOGY. Crown 8vo. [*In preparation.*
DEGENERATION: A CHAPTER IN DARWINISM. Illustrated. Crown 8vo. 2s. 6d. (*Nature Series.*)

Lubbock.—Works by SIR JOHN LUBBOCK, M.P., F.R.S., D.C.L.
THE ORIGIN AND METAMORPHOSES OF INSECTS. With numerous Illustrations. New Edition. Crown 8vo. 3s. 6d. (*Nature Series.*)
ON BRITISH WILD FLOWERS CONSIDERED IN RELATION TO INSECTS. With numerous Illustrations. New Edition. Crown 8vo. 4s. 6d. (*Nature Series*).

M'Kendrick.—OUTLINES OF PHYSIOLOGY IN ITS RELATIONS TO MAN. By J. G. M'KENDRICK, M.D., F.R.S.E. With Illustrations. Crown 8vo. 12s. 6d.

Martin and Moale.—ON THE DISSECTION OF VERTE-
BRATE ANIMALS. By Professor H. N. MARTIN and W. A.
MOALE. Crown 8vo. [*In preparation.*
(See also page 41.)

Miall.—STUDIES IN COMPARATIVE ANATOMY.
No. I.—The Skull of the Crocodile: a Manual for Students. By
L. C. MIALL, Professor of Biology in the Yorkshire College and
Curator of the Leeds Museum. 8vo. 2s. 6d.
No. II.—Anatomy of the Indian Elephant. By L. C. MIALL and
F. GREENWOOD. With Illustrations. 8vo. 5s.

Mivart.—Works by ST. GEORGE MIVART, F.R.S. Lecturer in
Comparative Anatomy at St. Mary's Hospital.
LESSONS IN ELEMENTARY ANATOMY. With upwards of
400 Illustrations. Fcap. 8vo. 6s. 6d.
THE COMMON FROG. With numerous Illustrations. Crown
8vo. 3s. 6d. (*Nature Series.*)

Müller.—THE FERTILISATION OF FLOWERS. By Pro-
fessor HERMANN MÜLLER. Translated and Edited by D'ARCY
W. THOMPSON, B.A., Scholar of Trinity College, Cambridge.
With a Preface by CHARLES DARWIN, F.R.S. With numerous
Illustrations. Medium 8vo. 21s.

Oliver.—Works by DANIEL OLIVER, F.R.S., &c., Professor of
Botany in University College, London, &c.
FIRST BOOK OF INDIAN BOTANY. With numerous Illus-
trations. Extra fcap. 8vo. 6s. 6d.
LESSONS IN ELEMENTARY BOTANY. With nearly 200
Illustrations. New Edition. Fcap. 8vo. 4s. 6d.

Parker.—A COURSE OF INSTRUCTION IN ZOOTOMY
(VERTEBRATA). By T. JEFFREY PARKER, B.Sc. London,
Professor of Biology in the University of Otago, New Zealand.
With Illustrations. Crown 8vo. 8s. 6d.

Parker and Bettany.—THE MORPHOLOGY OF THE
SKULL. By Professor PARKER and G. T. BETTANY. Illustrated. Crown 8vo. 10s. 6d.

Romanes.—THE SCIENTIFIC EVIDENCES OF ORGANIC
EVOLUTION. By G. J. ROMANES, M.A., LL.D., F.R.S.,
Zoological Secretary to the Linnean Society. Crown 8vo. 2s. 6d.
(*Nature Series.*)

Smith.—Works by JOHN SMITH, A.L.S., &c.
A DICTIONARY OF ECONOMIC PLANTS. Their History,
Products, and Uses. 8vo. 14s.

SCIENCE.

Smith.—Works by JOHN SMITH, A.L.S., &c. (*continued*)—
DOMESTIC BOTANY : An Exposition of the Structure and Classification of Plants, and their Uses for Food, Clothing, Medicine, and Manufacturing Purposes. With Illustrations. New Issue. Crown 8vo. 12*s.* 6*d.*

Smith (W. G.)—DISEASES OF FIELD AND GARDEN CROPS, CHIEFLY SUCH AS ARE CAUSED BY FUNGI. By WORTHING G. SMITH, F.L.S., M.A.I., Member of the Scientific Committee R.H.S. With 143 New Illustrations drawn and engraved from Nature by the Author. Fcap. 8vo. 4*s.* 6*d.*

MEDICINE.

Brunton.—Works by T. LAUDER BRUNTON, M.D., Sc.D., F.R.C.P., F.R.S., Examiner in Materia Medica in the University of London, late Examiner in Materia Medica in the University of Edinburgh, and the Royal College of Physicians, London.

A TREATISE ON MATERIA MEDICA. 8vo. [*In the press.*

TABLES OF MATERIA MEDICA : A Companion to the Materia Medica Museum. With Illustrations. New Edition Enlarged. 8vo. 10*s.* 6*d.*

Hamilton.—A TEXT-BOOK OF PATHOLOGY. By D. J. HAMILTON, Professor of Pathological Anatomy (Sir Erasmus Wilson Chair), University of Aberdeen. 8vo. [*In preparation.*

Ziegler-Macalister.—TEXT-BOOK OF PATHOLOGICAL ANATOMY AND PATHOGENESIS. By Professor ERNST ZIEGLER of Tübingen. Translated and Edited for English Students by DONALD MACALISTER, M.A., M.D., B.Sc., M.R.C.P., Fellow and Medical Lecturer of St. John's College, Cambridge, Physician to Addenbrooke's Hospital, and Teacher of Medicine in the University. With numerous Illustrations. Medium 8vo.
Part I.—GENERAL PATHOLOGICAL ANATOMY. 12*s.* 6*d.*
Part II.—SPECIAL PATHOLOGICAL ANATOMY. Sections I.—VIII. 12*s.* 6*d.* [PART III. *in preparation.*

ANTHROPOLOGY.

Flower.—FASHION IN DEFORMITY, as Illustrated in the Customs of Barbarous and Civilised Races. By Professor FLOWER, F.R.S., F.R.C.S. With Illustrations. Crown 8vo. 2*s.* 6*d.* (*Nature Series*).

Tylor.—ANTHROPOLOGY. An Introduction to the Study of Man and Civilisation. By E. B. TYLOR, D.C.L., F.R.S. With numerous Illustrations. Crown 8vo. 7s. 6d.

PHYSICAL GEOGRAPHY & GEOLOGY.

Blanford.—THE RUDIMENTS OF PHYSICAL GEOGRAPHY FOR THE USE OF INDIAN SCHOOLS; with a Glossary of Technical Terms employed. By H. F. BLANFORD, F.R.S. New Edition, with Illustrations. Globe 8vo. 2s. 6d.

Geikie.—Works by ARCHIBALD GEIKIE, F.R.S., Director General of the Geological Surveys of the United Kingdom.
PRIMER OF PHYSICAL GEOGRAPHY. With numerous Illustrations. New Edition. With Questions. 18mo. 1s. (*Science Primers.*)
ELEMENTARY LESSONS IN PHYSICAL GEOGRAPHY. With numerous Illustrations. New Edition. Fcap. 8vo. 4s. 6d.
QUESTIONS ON THE SAME. 1s. 6d.
PRIMER OF GEOLOGY. With numerous Illustrations. New Edition. 18mo. 1s. (*Science Primers.*)
ELEMENTARY LESSONS IN GEOLOGY. With Illustrations. Fcap. 8vo. [*In preparation.*
TEXT-BOOK OF GEOLOGY. With numerous Illustrations. 8vo. 28s.
OUTLINES OF FIELD GEOLOGY. With Illustrations. New Edition. Extra fcap. 8vo. 3s. 6d.

Huxley.—PHYSIOGRAPHY. An Introduction to the Study of Nature. By Professor HUXLEY, P.R.S. With numerous Illustrations, and Coloured Plates. New and Cheaper Edition. Crown 8vo. 6s.

Phillips.—A TREATISE ON ORE DEPOSITS. By J. ARTHUR PHILLIPS, F.R.S., V.P.G.S., F.C.S., M.Inst.C.E., Ancien Élève de l'École des Mines, Paris; Author of "A Manual of Metallurgy," "The Mining and Metallurgy of Gold and Silver," &c. With numerous Illustrations. 8vo. 25s.

AGRICULTURE.

Frankland.—AGRICULTURAL CHEMICAL ANALYSIS, A Handbook of. By PERCY FARADAY FRANKLAND, Ph.D., B.Sc., F.C.S., Associate of the Royal School of Mines, and Demonstrator of Practical and Agricultural Chemistry in the Normal School of Science and Royal School of Mines, South Kensington Museum. Founded upon *Leitfaden für die Agricultur Chemiche Analyse*, von Dr. F. KROCKER. Crown 8vo. 7s. 6d.

Smith (Worthington G.).—DISEASES OF FIELD AND GARDEN CROPS, CHIEFLY SUCH AS ARE CAUSED BY FUNGI. By WORTHINGTON G. SMITH, F.L.S., M.A.I., Member of the Scientific Committee of the R.H.S. With 143 Illustrations, drawn and engraved from Nature by the Author. Fcap. 8vo. 4s. 6d.

Tanner.—Works by HENRY TANNER, F.C.S., M.R.A.C., Examiner in the Principles of Agriculture under the Government Department of Science; Director of Education in the Institute of Agriculture, South Kensington, London; sometime Professor of Agricultural Science, University College, Aberystwith.

ELEMENTARY LESSONS IN THE SCIENCE OF AGRICULTURAL PRACTICE. Fcap. 8vo. 3s. 6d.

FIRST PRINCIPLES OF AGRICULTURE. 18mo. 1s.

THE PRINCIPLES OF AGRICULTURE. A Series of Reading Books for use in Elementary Schools. Prepared by HENRY TANNER, F.C.S., M.R.A.C. Extra fcap. 8vo.
  I. The Alphabet of the Principles of Agriculture. 6d.
  II. Further Steps in the Principles of Agriculture. 1s.
  III. Elementary School Readings on the Principles of Agriculture for the third stage. 1s.

POLITICAL ECONOMY.

Cossa.—GUIDE TO THE STUDY OF POLITICAL ECONOMY. By Dr. LUIGI COSSA, Professor in the University of Pavia. Translated from the Second Italian Edition. With a Preface by W. STANLEY JEVONS, F.R.S. Crown 8vo. 4s. 6d.

Fawcett (Mrs.).—Works by MILLICENT GARRETT FAWCETT:—
POLITICAL ECONOMY FOR BEGINNERS, WITH QUESTIONS. Fourth Edition. 18mo. 2s. 6d.
TALES IN POLITICAL ECONOMY. Crown 8vo. 3s.

Fawcett.—A MANUAL OF POLITICAL ECONOMY. By Right Hon. HENRY FAWCETT, M.P., F.R.S. Sixth Edition, revised, with a chapter on "State Socialism and the Nationalisation of the Land," and an Index. Crown 8vo. 12s.

Jevons.—PRIMER OF POLITICAL ECONOMY. By W. STANLEY JEVONS, LL.D., M.A., F.R.S. New Edition. 18mo. 1s. (*Science Primers.*)

Marshall.—THE ECONOMICS OF INDUSTRY. By A. MARSHALL, M.A., late Principal of University College, Bristol, and MARY P. MARSHALL, late Lecturer at Newnham Hall, Cambridge. Extra fcap. 8vo. 2s. 6d.

Sidgwick.—THE PRINCIPLES OF POLITICAL ECONOMY By Professor HENRY SIDGWICK, M.A., LL.D. Knightbride. Professor of Moral Philosophy in the University of Cambridge, &c., Author of "The Methods of Ethics." 8vo. 16s.

48   MACMILLAN'S EDUCATIONAL CATALOGUE.

Walker.—POLITICAL ECONOMY. By FRANCIS A. WALKER, M.A., Ph.D., Author of "The Wages Question," "Money," "Money in its Relation to Trade," &c. 8vo. 10s. 6d.

MENTAL & MORAL PHILOSOPHY.

Caird.—MORAL PHILOSOPHY, An Elementary Treatise on. By Prof. E. CAIRD, of Glasgow University. Fcap. 8vo.
[*In preparation.*

Calderwood.—HANDBOOK OF MORAL PHILOSOPHY. By the Rev. HENRY CALDERWOOD, LL.D., Professor of Moral Philosophy, University of Edinburgh. New Edition. Crown 8vo. 6s.

Clifford.—SEEING AND THINKING. By the late Professor W. K. CLIFFORD, F.R.S. With Diagrams. Crown 8vo. 3s. 6d. (*Nature Series.*)

Jardine.—THE ELEMENTS OF THE PSYCHOLOGY OF COGNITION. By the Rev. ROBERT JARDINE, B.D., D.Sc. (Edin.), Ex-Principal of the General Assembly's College, Calcutta. Second Edition, revised and improved. Crown 8vo. 6s. 6d.

Jevons.—Works by the late W. STANLEY JEVONS, LL.D., M.A., F.R.S.
PRIMER OF LOGIC. New Edition. 18mo. 1s. (*Science Primers.*)
ELEMENTARY LESSONS IN LOGIC; Deductive and Inductive, with copious Questions and Examples, and a Vocabulary of Logical Terms. New Edition. Fcap. 8vo. 3s. 6d.
THE PRINCIPLES OF SCIENCE. A Treatise on Logic and Scientific Method. New and Revised Edition. Crown 8vo. 12s. 6d.
STUDIES IN DEDUCTIVE LOGIC. Second Edition. Crown 8vo. 6s.

Keynes.—FORMAL LOGIC, Studies and Exercises in. Including a Generalisation of Logical Processes in their application to Complex Inferences. By JOHN NEVILLE KEYNES, M.A., late Fellow of Pembroke College, Cambridge. Crown 8vo. 10s. 6d.

Robertson.—ELEMENTARY LESSONS IN PSYCHOLOGY. By G. CROOM ROBERTSON, Professor of Mental Philosophy, &c., University College, London. [*In preparation.*

Sidgwick.—THE METHODS OF ETHICS. By Professor HENRY SIDGWICK, M.A., LL.D. Cambridge, &c. Third Edition. 8vo. 14s. A Supplement to the Second Edition, containing all the important Additions and Alterations in the Third Edition. Demy 8vo. 6s.

HISTORY AND GEOGRAPHY.

Arnold.—THE ROMAN SYSTEM OF PROVINCIAL ADMINISTRATION TO THE ACCESSION OF CONSTANTINE THE GREAT. By W. T. ARNOLD, B.A. Crown 8vo. 6s.

"Ought to prove a valuable handbook to the student of Roman history."—GUARDIAN.

Beesly.—STORIES FROM THE HISTORY OF ROME. By Mrs. BEESLY. Fcap. 8vo. 2s. 6d.

"The attempt appears to us in every way successful. The stories are interesting in themselves, and are told with perfect simplicity and good feeling." — DAILY NEWS.

Bryce.—THE HOLY ROMAN EMPIRE. By JAMES BRYCE, D.C.L., Fellow of Oriel College, and Regius Professor of Civil Law in the University of Oxford. Seventh Edition. Crown 8vo. 7s. 6d.

Brook.—FRENCH HISTORY FOR ENGLISH CHILDREN. By SARAH BROOK. With Coloured Maps. Crown 8vo. 6s.

Clarke.—CLASS-BOOK OF GEOGRAPHY. By C. B. CLARKE, M.A., F.L.S., F.G.S., F.R.S. New Edition, with Eighteen Coloured Maps. Fcap. 8vo. 3s.

Freeman.—OLD ENGLISH HISTORY. By Prof. EDWARD A. FREEMAN, D.C.L., LL.D., late Fellow of Trinity College, Oxford. With Five Coloured Maps. New Edition. Extra fcap. 8vo. 6s.

Fyffe.—A SCHOOL HISTORY OF GREECE. By C. A. FYFFE, M.A., Fellow of University College, Oxford. Crown 8vo. [*In preparation.*

Green. — Works by JOHN RICHARD GREEN, M.A., LL.D., late Honorary Fellow of Jesus College, Oxford.

SHORT HISTORY OF THE ENGLISH PEOPLE. With Coloured Maps, Genealogical Tables, and Chronological Annals. Crown 8vo. 8s. 6d. 102nd Thousand.

"Stands alone as the one general history of the country, for the sake of which all others, if young and old are wise, will be speedily and surely set aside."—ACADEMY.

ANALYSIS OF ENGLISH HISTORY, based on Green's "Short History of the English People." By C. W. A. TAIT, M.A., Assistant-Master, Clifton College. Crown 8vo. 3s. 6d.

READINGS FROM ENGLISH HISTORY. Selected and Edited by JOHN RICHARD GREEN. Three Parts. Globe 8vo. 1s. 6d. each. I. Hengist to Cressy. II. Cressy to Cromwell. III. Cromwell to Balaklava.

A SHORT GEOGRAPHY OF THE BRITISH ISLANDS. By JOHN RICHARD GREEN and ALICE STOPFORD GREEN. With Maps. Fcap. 8vo. 3s. 6d.

Grove.—A PRIMER OF GEOGRAPHY. By Sir GEORGE GROVE, D.C.L., F.R.G.S. With Illustrations. 18mo. 1s. (*Science Primers.*)

Guest.—LECTURES ON THE HISTORY OF ENGLAND. By M. J. GUEST. With Maps. Crown 8vo. 6s.

"It is not too much to assert that this is one of the very best class books of English History for young students ever published."—SCOTSMAN.

Historical Course for Schools—Edited by EDWARD A. FREEMAN, D.C.L., late Fellow of Trinity College, Oxford, Regius Professor of Modern History in the University of Oxford.

I.—GENERAL SKETCH OF EUROPEAN HISTORY. By Prof. EDWARD A. FREEMAN, D.C.L. New Edition, revised and enlarged, with Chronological Table, Maps, and Index. 18mo. 3s. 6d.

II.—HISTORY OF ENGLAND. By EDITH THOMPSON. New Edition, revised and enlarged, with Coloured Maps. 18mo. 2s. 6d.

III.—HISTORY OF SCOTLAND. By MARGARET MACARTHUR. New Edition. 18mo. 2s.

IV.—HISTORY OF ITALY. By the Rev. W. HUNT, M.A. New Edition, with Coloured Maps. 18mo. 3s. 6d.

V.—HISTORY OF GERMANY. By J. SIME, M.A. 18mo. 3s.

VI.—HISTORY OF AMERICA. By JOHN A. DOYLE. With Maps. 18mo. 4s. 6d.

VII.—EUROPEAN COLONIES. By E. J. PAYNE, M.A. With Maps. 18mo. 4s. 6d.

VIII.—FRANCE. By CHARLOTTE M. YONGE. With Maps. 18mo. 3s. 6d.

GREECE. By Prof. EDWARD A. FREEMAN, D.C.L.
[*In preparation.*

ROME. By Prof. EDWARD A. FREEMAN, D.C.L. [*In preparation.*

History Primers—Edited by JOHN RICHARD GREEN, M.A., LL.D., Author of "A Short History of the English People."

ROME. By the Rev. M. CREIGHTON, M.A.; late Fellow and Tutor of Merton College, Oxford. With Eleven Maps. 18mo. 1s.

"The author has been curiously successful in telling in an intelligent way the story of Rome from first to last."—SCHOOL BOARD CHRONICLE.

GREECE. By C. A. FYFFE, M.A., Fellow and late Tutor of University College, Oxford. With Five Maps. 18mo. 1s.

"We give our unqualified praise to this little manual.'—SCHOOLMASTER.

History and Geography.

History Primers *Continued—*

EUROPEAN HISTORY. By Prof. E. A. FREEMAN, D.C.L., LL.D. With Maps. 18mo. 1s.

"The work is always clear, and forms a luminous key to European history."—SCHOOL BOARD CHRONICLE.

GREEK ANTIQUITIES. By the Rev. J. P. MAHAFFY, M.A. Illustrated. 18mo. 1s.

"All that is necessary for the scholar to know is told so compactly yet so fully, and in a style so interesting, that it is impossible for even the dullest boy to look on this little work in the same light as he regards his other school books."—SCHOOLMASTER.

CLASSICAL GEOGRAPHY. By H. F. TOZER, M.A. 18mo. 1s.

"Another valuable aid to the study of the ancient world. . . . It contains an enormous quantity of information packed into a small space, and at the same time communicated in a very readable shape."—JOHN BULL.

GEOGRAPHY. By Sir GEORGE GROVE, D.C.L. With Maps. 18mo. 1s.

"A model of what such a work should be. . . . We know of no short treatise better suited to infuse life and spirit into the dull lists of proper names of which our ordinary class-books so often almost exclusively consist."—TIMES.

ROMAN ANTIQUITIES. By Professor WILKINS. Illustrated. 18mo. 1s.

"A little book that throws a blaze of light on Roman history, and is, moreover intensely interesting."—SCHOOL BOARD CHRONICLE.

FRANCE. By CHARLOTTE M. YONGE. 18mo. 1s.

"May be considered a wonderfully successful piece of work. . . . Its general merit as a vigorous and clear sketch, giving in a small space a vivid idea of the history of France, remains undeniable."—SATURDAY REVIEW.

Hole.—A GENEALOGICAL STEMMA OF THE KINGS OF ENGLAND AND FRANCE. By the Rev. C. HOLE. On Sheet. 1s.

Jennings.—CHRONOLOGICAL TABLES. Compiled by Rev. A. C. JENNINGS. [*In the press.*

Kiepert—A MANUAL OF ANCIENT GEOGRAPHY. From the German of Dr. H. KIEPERT. Crown 8vo. 5s.

Lethbridge.—A SHORT MANUAL OF THE HISTORY OF INDIA. With an Account of INDIA AS IT IS. The Soil, Climate, and Productions; the People, their Races, Religions, Public Works, and Industries; the Civil Services, and System of Administration. By ROPER LETHBRIDGE, M.A., C.I.E., late Scholar of Exeter College, Oxford, formerly Principal of Kishnaghur College, Bengal, Fellow and sometime Examiner of the Calcutta University. With Maps. Crown 8vo. 5s.

Michelet.—A SUMMARY OF MODERN HISTORY. Translated from the French of M. MICHELET, and continued to the Present Time, by M. C. M. SIMPSON. Globe 8vo. 4s. 6d.

Otté.—SCANDINAVIAN HISTORY. By E. C. OTTÉ. With Maps. Globe 8vo. 6s.

Ramsay.—A SCHOOL HISTORY OF ROME. By G. G. RAMSAY, M.A., Professor of Humanity in the University of Glasgow. With Maps. Crown 8vo. [*In preparation.*

Tait.—ANALYSIS OF ENGLISH HISTORY, based on Green's "Short History of the English People." By C. W. A. TAIT, M.A., Assistant-Master, Clifton College. Crown 8vo. 3s. 6d.

Wheeler.—A SHORT HISTORY OF INDIA AND OF THE FRONTIER STATES OF AFGHANISTAN, NEPAUL, AND BURMA. By J. TALBOYS WHEELER. With Maps. Crown 8vo. 12s.

"It is the best book of the kind we have ever seen, and we recommend it to a place in every school library."—EDUCATIONAL TIMES.

Yonge (Charlotte M.).—A PARALLEL HISTORY OF FRANCE AND ENGLAND: consisting of Outlines and Dates. By CHARLOTTE M. YONGE, Author of "The Heir of Redclyffe," &c., &c. Oblong 4to. 3s. 6d.

CAMEOS FROM ENGLISH HISTORY.—FROM ROLLO TO EDWARD II. By the Author of "The Heir of Redclyffe.' Extra fcap. 8vo. New Edition. 5s.

A SECOND SERIES OF CAMEOS FROM ENGLISH HISTORY. — THE WARS IN FRANCE. New Edition. Extra fcap. 8vo. 5s.

A THIRD SERIES OF CAMEOS FROM ENGLISH HISTORY. —THE WARS OF THE ROSES. New Edition. Extra fcap. 8vo. 5s.

CAMEOS FROM ENGLISH HISTORY—A FOURTH SERIES. REFORMATION TIMES. Extra fcap. 8vo. 5s.

CAMEOS FROM ENGLISH HISTORY.—A FIFTH SERIES. ENGLAND AND SPAIN. Extra fcap. 8vo. 5s.

EUROPEAN HISTORY. Narrated in a Series of Historical Selections from the Best Authorities. Edited and arranged by E. M. SEWELL and C. M. YONGE. First Series, 1003—1154. New Edition. Crown 8vo. 6s. Second Series, 1088—1228. New Edition. Crown 8vo. 6s.

MODERN LANGUAGES AND LITERATURE.

(1) English, (2) French, (3) German, (4) Modern Greek, (5) Italian.

ENGLISH.

Abbott.—A SHAKESPEARIAN GRAMMAR. An attempt to illustrate some of the Differences between Elizabethan and Modern English. By the Rev. E. A. ABBOTT, D.D., Head Master of the City of London School. New Edition. Extra fcap. 8vo. 6s.

Brooke.—PRIMER OF ENGLISH LITERATURE. By the Rev. STOPFORD A. BROOKE, M.A. 18mo. 1s. (*Literature Primers.*)

Butler.—HUDIBRAS. Edited, with Introduction and Notes, by ALFRED MILNES, M.A. Lon., late Fellow of Lincoln College, Oxford. Extra fcap 8vo. Part I. 3s. 6d. Parts II. and III. 4s. 6d.

Cowper's TASK: AN EPISTLE TO JOSEPH HILL, ESQ.; TIROCINIUM, or a Review of the Schools; and THE HISTORY OF JOHN GILPIN. Edited, with Notes, by WILLIAM BENHAM, B.D. Globe 8vo. 1s. (*Globe Readings from Standard Authors.*)

Dowden.—SHAKESPEARE. By Professor DOWDEN. 18mo. 1s. (*Literature Primers.*)

Dryden.—SELECT PROSE WORKS. Edited, with Introduction and Notes, by Professor C. D. YONGE. Fcap. 8vo. 2s. 6d.

Gladstone.—SPELLING REFORM FROM AN EDUCATIONAL POINT OF VIEW. By J. H. GLADSTONE, Ph.D., F.R.S., Member of the School Board for London. New Edition. Crown 8vo. 1s. 6d.

Globe Readers. For Standards I.—VI. Edited by A. F. MURISON. Sometime English Master at the Aberdeen Grammar School. With Illustrations. Globe 8vo.

Primer I. (48 pp.) 3d.	Book III. (232 pp.) 1s. 3d.
Primer II. (48 pp.) 3d.	Book IV. (328 pp.) 1s. 9d.
Book I. (96 pp.) 6d.	Book V. (416 pp.) 2s.
Book II. (136 pp.) 9d.	Book VI. (448 pp.) 2s. 6d.

"Among the numerous sets of readers before the public the present series is honourably distinguished by the marked superiority of its materials and the careful ability with which they have been adapted to the growing capacity of the pupils. The plan of the two primers is excellent for facilitating the child's first attempts to read. In the first three following books there is abundance of entertaining reading. . . . Better food for young minds could hardly be found."—THE ATHENÆUM.

*The Shorter Globe Readers.—With Illustrations. Globe 8vo.

Primer I. (48 pp.) 3d.	Standard III. (178 pp.) 1s.
Primer II. (48 pp.) 3d.	Standard IV. (182 pp.) 1s.
Standard I. (92 pp.) 6d.	Standard V. (216 pp.) 1s. 3d.
Standard II. (124 pp.) 9d.	Standard VI. (228 pp.) 1s. 6d.

* This Series has been abridged from "The Globe Readers" to meet the demand for smaller reading books.

GLOBE READINGS FROM STANDARD AUTHORS.

Cowper's TASK: AN EPISTLE TO JOSEPH HILL, ESQ.; TIROCINIUM, or a Review of the Schools; and THE HISTORY OF JOHN GILPIN. Edited, with Notes, by WILLIAM BENHAM, B.D. Globe 8vo. 1s.

Goldsmith's VICAR OF WAKEFIELD. With a Memoir of Goldsmith by Professor MASSON. Globe 8vo. 1s.

Lamb's (Charles) TALES FROM SHAKESPEARE Edited, with Preface, by ALFRED AINGER, M.A. Globe 8vo. 2s.

Scott's (Sir Walter) LAY OF THE LAST MINSTREL; and THE LADY OF THE LAKE. Edited, with Introductions and Notes, by FRANCIS TURNER PALGRAVE. Globe 8vo. 1s.

MARMION; and the LORD OF THE ISLES. By the same Editor. Globe 8vo. 1s.

The Children's Garland from the Best Poets.—Selected and arranged by COVENTRY PATMORE. Globe 8vo. 2s.

Yonge (Charlotte M.).—A BOOK OF GOLDEN DEEDS OF ALL TIMES AND ALL COUNTRIES. Gathered and narrated anew by CHARLOTTE M. YONGE, the Author of "The Heir of Redclyffe." Globe 8vo. 2s.

Goldsmith.—THE TRAVELLER, or a Prospect of Society; and THE DESERTED VILLAGE. By OLIVER GOLDSMITH. With Notes, Philological and Explanatory, by J. W. HALES, M.A. Crown 8vo. 6d.

THE VICAR OF WAKEFIELD. With a Memoir of Goldsmith by Professor MASSON. Globe 8vo. 1s. (*Globe Readings from Standard Authors.*)

SELECT ESSAYS. Edited, with Introduction and Notes, by Professor C. D. YONGE. Fcap. 8vo. 2s. 6d.

MODERN LANGUAGES AND LITERATURE. 55

Hales.—LONGER ENGLISH POEMS, with Notes, Philological and Explanatory, and an Introduction on the Teaching of English. Chiefly for Use in Schools. Edited by J. W. HALES, M.A., Professor of English Literature at King's College, London. New Edition. Extra fcap. 8vo. 4s. 6d.

Johnson's LIVES OF THE POETS. The Six Chief Lives (Milton, Dryden, Swift, Addison, Pope, Gray), with Macaulay's "Life of Johnson." Edited with Preface by MATTHEW ARNOLD. Crown 8vo. 6s.

Lamb (Charles).—TALES FROM SHAKESPEARE. Edited, with Preface, by ALFRED AINGER, M.A. Globe 8vo. 2s. (*Globe Readings from Standard Authors.*)

Literature Primers—Edited by JOHN RICHARD GREEN, M.A., LL.D., Author of "A Short History of the English People."
ENGLISH COMPOSITION. By Professor NICHOL. 18mo. 1s.
ENGLISH GRAMMAR. By the Rev. R. MORRIS, LL.D., sometime President of the Philological Society. 18mo, cloth. 1s.
ENGLISH GRAMMAR EXERCISES. By R. MORRIS, LL.D., and H. C. BOWEN, M.A. 18mo. 1s.
EXERCISES ON MORRIS'S PRIMER OF ENGLISH GRAMMAR. By JOHN WETHERELL, of the Middle School, Liverpool College. 18mo. 1s.
ENGLISH LITERATURE. By STOPFORD BROOKE, M.A. New Edition. 18mo. 1s.
SHAKSPERE. By Professor DOWDEN. 18mo. 1s.
THE CHILDREN'S TREASURY OF LYRICAL POETRY. Selected and arranged with Notes by FRANCIS TURNER PALGRAVE. In Two Parts. 18mo. 1s. each.
PHILOLOGY. By J. PEILE, M.A. 18mo. 1s.

In preparation:—
HISTORY OF THE ENGLISH LANGUAGE. By J. A. H. MURRAY, LL.D.
SPECIMENS OF THE ENGLISH LANGUAGE. To Illustrate the above. By the same Author.

Macmillan's Reading Books.—Adapted to the English and Scotch Codes. Bound in Cloth.
PRIMER. 18mo. (48 pp.) 2d.
BOOK I. for Standard I. 18mo. (96 pp.) 4d.
„ II. „ II. 18mo. (144 pp.) 5d.
„ III. „ III. 18mo. (160 pp.) 6d.
„ IV. „ IV. 18mo. (176 pp.) 8d.
„ V. „ V. 18mo. (380 pp.) 1s.

Macmillan's Reading-Books *Continued*—

BOOK VI. for Standard VI. Crown 8vo. (430 pp.) 2s.

Book VI. is fitted for higher Classes, and as an Introduction to English Literature.

"They are far above any others that have appeared both in form and substance. . . The editor of the present series has rightly seen that reading books must aim chiefly at giving to the pupils the power of accurate, and, if possible, apt and skilful expression; at cultivating in them a good literary taste, and at arousing a desire of further reading.' This is done by taking care to select the extracts from true English classics, going up in Standard VI. course to Chaucer, Hooker, and Bacon, as well as Wordsworth, Macaulay, and Froude. . . . This is quite on the right track, and indicates justly the ideal which we ought to set before us."— GUARDIAN.

Macmillan's Copy-Books—

Published in two sizes, viz. :—
    1. Large Post 4to. Price 4*d*. each.
    2. Post Oblong. Price 2*d*. each.

  1. INITIATORY EXERCISES AND SHORT LETTERS.
  2. WORDS CONSISTING OF SHORT LETTERS.
*3. LONG LETTERS. With words containing Long Letters—Figures.
*4. WORDS CONTAINING LONG LETTERS.
4a. PRACTISING AND REVISING COPY-BOOK. For Nos. 1 to 4.
*5. CAPITALS AND SHORT HALF-TEXT. Words beginning with a Capital.
*6. HALF-TEXT WORDS beginning with Capitals—Figures.
*7. SMALL-HAND AND HALF-TEXT. With Capitals and Figures.
*8. SMALL-HAND AND HALF-TEXT. With Capitals and Figures.
8a. PRACTISING AND REVISING COPY-BOOK. For Nos. 5 to 8.
*9. SMALL-HAND SINGLE HEADLINES—Figures.
10. SMALL-HAND SINGLE HEADLINES—Figures.
11. SMALL-HAND DOUBLE HEADLINES—Figures.
12. COMMERCIAL AND ARITHMETICAL EXAMPLES, &c.
12a. PRACTISING AND REVISING COPY-BOOK. For Nos. 8 to 12.

    * *These numbers may be had with Goodman's Patent Sliding Copies.* Large Post 4to. Price 6*d*. each.

MODERN LANGUAGES AND LITERATURE. 57

Martin.—THE POET'S HOUR : Poetry selected and arranged for Children. By FRANCES MARTIN. New Edition. 18mo. 2s. 6d.

SPRING-TIME WITH THE POETS : Poetry selected by FRANCES MARTIN. New Edition. 18mo. 3s. 6d.

Milton.—By STOPFORD BROOKE, M.A. Fcap. 8vo. 1s. 6d. (*Classical Writers Series.*)

Morris.—Works by the Rev. R. MORRIS, LL.D.

HISTORICAL OUTLINES OF ENGLISH ACCIDENCE, comprising Chapters on the History and Development of the Language, and on Word-formation. New Edition. Extra fcap. 8vo. 6s.

ELEMENTARY LESSONS IN HISTORICAL ENGLISH GRAMMAR, containing Accidence and Word-formation. New Edition. 18mo. 2s. 6d.

PRIMER OF ENGLISH GRAMMAR. 18mo. 1s. (See also *Literature Primers.*)

Oliphant.—THE OLD AND MIDDLE ENGLISH. A New Edition of "THE SOURCES OF STANDARD ENGLISH," revised and greatly enlarged. By T. L. KINGTON OLIPHANT. Extra fcap. 8vo. 9s.

Palgrave.—THE CHILDREN'S TREASURY OF LYRICAL POETRY. Selected and arranged, with Notes, by FRANCIS TURNER PALGRAVE. 18mo. 2s. 6d. Also in Two Parts. 18mo. 1s. each.

Patmore.—THE CHILDREN'S GARLAND FROM THE BEST POET'S. Selected and arranged by COVENTRY PATMORE. Globe 8vo. 2s. (*Globe Readings from Standard Authors.*)

Plutarch.—Being a Selection from the Lives which Illustrate Shakespeare. North's Translation. Edited, with Introductions, Notes, Index of Names, and Glossarial Index, by the Rev. W. W. SKEAT, M.A. Crown 8vo. 6s.

Scott's (Sir Walter) LAY OF THE LAST MINSTREL, and THE LADY OF THE LAKE. Edited, with Introduction and Notes, by FRANCIS TURNER PALGRAVE. Globe 8vo. 1s. (*Globe Readings from Standard Authors.*)

MARMION ; and THE LORD OF THE ISLES. By the same Editor. Globe 8vo. 1s. (*Globe Readings from Standard Authors.*)

Shakespeare.—A SHAKESPEARE MANUAL. By F. G. FLEAY, M.A., late Head Master of Skipton Grammar School. Second Edition. Extra fcap. 8vo. 4s. 6d.

AN ATTEMPT TO DETERMINE THE CHRONOLOGICAL ORDER OF SHAKESPEARE'S PLAYS. By the Rev. H. PAINE STOKES, B.A. Extra fcap. 8vo. 4s. 6d.

THE TEMPEST. With Glossarial and Explanatory Notes. By the Rev. J. M. JEPHSON. New Edition. 18mo. 1s.

PRIMER OF SHAKESPEARE. By Professor DOWDEN. 18mo. 1s. (*Literature Primers.*)

Sonnenschein and Meiklejohn.— THE ENGLISH METHOD OF TEACHING TO READ. By A. SONNENSCHEIN and J. M. D. MEIKLEJOHN, M.A. Fcap. 8vo.

COMPRISING:

THE NURSERY BOOK, containing all the Two-Letter Words in the Language. 1d. (Also in Large Type on Sheets for School Walls. 5s.)

THE FIRST COURSE, consisting of Short Vowels with Single Consonants. 6d.

THE SECOND COURSE, with Combinations and Bridges, consisting of Short Vowels with Double Consonants. 6d.

THE THIRD AND FOURTH COURSES, consisting of Long Vowels, and all the Double Vowels in the Language. 6d.

"These are admirable books, because they are constructed on a principle, and that the simplest principle on which it is possible to learn to read English."—SPECTATOR.

Taylor.—WORDS AND PLACES; or, Etymological Illustrations of History, Ethnology, and Geography. By the Rev. ISAAC TAYLOR, M.A. Third and Cheaper Edition, revised and compressed. With Maps. Globe 8vo. 6s.

Tennyson.—The COLLECTED WORKS of ALFRED, LORD TENNYSON, Poet Laureate. An Edition for Schools. In Four Parts. Crown 8vo. 2s. 6d. each.

Thring.—THE ELEMENTS OF GRAMMAR TAUGHT IN ENGLISH. By EDWARD THRING, M.A., Head Master of Uppingham. With Questions. Fourth Edition. 18mo. 2s.

Trench (Archbishop).—Works by R. C. TRENCH, D.D., Archbishop of Dublin.

HOUSEHOLD BOOK OF ENGLISH POETRY, Selected and Arranged, with Notes. Third Edition. Extra fcap. 8vo. 5s. 6d.

Trench (Archbishop) Works by, *continued*—
 ON THE STUDY OF WORDS. Seventeenth Edition, revised. Fcap. 8vo. 5s.
 ENGLISH, PAST AND PRESENT. Eleventh Edition, revised and improved. Fcap. 8vo. 5s.
 A SELECT GLOSSARY OF ENGLISH WORDS, used formerly in Senses Different from their Present. Fifth Edition, revised and enlarged. Fcap. 8vo. 5s.

Vaughan (C.M.).—WORDS FROM THE POETS. By C. M. VAUGHAN. New Edition. 18mo, cloth. 1s.

Ward.—THE ENGLISH POETS. Selections, with Critical Introductions by various Writers and a General Introduction by MATTHEW ARNOLD. Edited by T. H. WARD, M.A. 4 Vols. Vol. I. CHAUCER TO DONNE.—Vol. II. BEN JONSON TO DRYDEN.—Vol. III. ADDISON TO BLAKE.—Vol. IV. WORDSWORTH TO ROSSETTI. Crown 8vo. Each 7s. 6d.

Wetherell.—EXERCISES ON MORRIS'S PRIMER OF ENGLISH GRAMMAR. By JOHN WETHERELL, M.A. 18mo. 1s. (*Literature Primers.*)

Wrightson.—THE FUNCTIONAL ELEMENTS OF AN ENGLISH SENTENCE, an Examination of. Together with a New System of Analytical Marks. By the Rev. W. G. WRIGHTSON, M.A., Cantab. Crown 8vo. 5s.

Yonge (Charlotte M.).—THE ABRIDGED BOOK OF GOLDEN DEEDS. A Reading Book for Schools and general readers. By the Author of "The Heir of Redclyffe." 18mo, cloth. 1s.
 GLOBE READINGS EDITION. Complete Edition. Globe 8vo. 2s. (See p. 53.)

FRENCH.

Beaumarchais.—LE BARBIER DE SEVILLE. Edited, with Introduction and Notes, by L. P. BLOUET, Assistant Master in St. Paul's School. Fcap. 8vo. 3s. 6d.

Bowen.—FIRST LESSONS IN FRENCH. By H. COURTHOPE BOWEN, M.A., Principal of the Finsbury Training College for Higher and Middle Schools. Extra fcap. 8vo. 1s.

Breymann.—Works by HERMANN BREYMANN, Ph.D., Professor of Philology in the University of Munich.
 A FRENCH GRAMMAR BASED ON PHILOLOGICAL PRINCIPLES. Second Edition. Extra fcap. 8vo. 4s. 6d.

Breymann—Works by HERMANN BREYMANN, Ph.D. (*continued*)
FIRST FRENCH EXERCISE BOOK. Extra fcap. 8vo. 4s. 6d.
SECOND FRENCH EXERCISE BOOK. Extra fcap. 8vo. 2s. 6d.

Fasnacht.—THE ORGANIC METHOD OF STUDYING LANGUAGES. By G. EUGÈNE FASNACHT, Author of "Macmillan's Progressive French Course," Editor of "Macmillan's Foreign School Classics," &c. Extra fcap. 8vo. I. French. 3s. 6d.

A SYNTHETIC FRENCH GRAMMAR FOR SCHOOLS By the same Author. Crown 8vo. 3s. 6d.

GRAMMAR AND GLOSSARY OF THE FRENCH LANGUAGE OF THE SEVENTEENTH CENTURY. By the same Author. Crown 8vo. [*In preparation.*

Macmillan's Primary Series of French and German Reading Books.—Edited by G. EUGÈNE FASNACHT, Assistant-Master in Westminster School. With Illustrations. Globe 8vo.

PERRAULT—CONTES DE FÉES. Edited, with Introduction, Notes, and Vocabulary, by G. E. FASNACHT. 1s.
LA FONTAINE—SELECT FABLES. Edited, with Introduction, Notes, and Vocabulary, by L. M. MORIARTY, M.A., Assistant-Master at Rossall. [*In preparation.*
GRIMM—HAUSMÄRCHEN. Edited, with Introduction, Notes, and Vocabulary, by G. E. FASNACHT. [*In preparation.*
G. SCHWAB—ODYSSEUS. With Introduction, Notes, and Vocabulary, by the same Editor. [*In preparation.*

Macmillan's Progressive French Course.—By G. EUGÈNE FASNACHT, Assistant-Master in Westminster School.
- I.—FIRST YEAR, containing Easy Lessons on the Regular Accidence. Extra fcap. 8vo. 1s.
- II.—SECOND YEAR, containing an Elementary Grammar with copious Exercises, Notes, and Vocabularies. A new Edition, enlarged and thoroughly revised. Extra fcap. 8vo. 2s.
- III.—THIRD YEAR, containing a Systematic Syntax, and Lessons in Composition. Extra fcap. 8vo. 2s. 6d.

THE TEACHER'S COMPANION TO MACMILLAN'S PROGRESSIVE FRENCH COURSE. Third Year. With Copious Notes, Hints for Different Renderings, Synonyms, Philological Remarks, &c. By G. E. FASNACHT. Globe 8vo. 4s. 6d.

THE TEACHER'S COMPANION TO MACMILLAN'S PROGRESSIVE FRENCH COURSE. Second Year.
[*In the press.*

MODERN LANGUAGES AND LITERATURE. 61

Macmillan's Progressive French Readers.—By G. EUGÈNE FASNACHT.

I.—FIRST YEAR, containing Fables, Historical Extracts, Letters, Dialogues, Fables, Ballads, Nursery Songs, &c., with Two Vocabularies: (1) in the order of subjects; (2) in alphabetical order. Extra fcap. 8vo. 2s. 6d.

II.—SECOND YEAR, containing Fiction in Prose and Verse, Historical and Descriptive Extracts, Essays, Letters, Dialogues, &c. Extra fcap. 8vo. 2s. 6d.

Macmillan's Foreign School Classics.—Edited by G. EUGÈNE FASNACHT. 18mo.

FRENCH.

CORNEILLE—LE CID. Edited by G. E. FASNACHT. 1s.

DUMAS—LES DEMOISELLES DE ST. CYR. Edited by VICTOR OGER, Lecturer in University College, Liverpool. [*In preparation.*

MOLIÈRE—LES FEMMES SAVANTES. By G. E. FASNACHT. 1s.

MOLIÈRE—LE MISANTHROPE. By the same Editor. 1s.

MOLIÈRE—LE MÉDECIN MALGRÉ LUI. By the same Editor. 1s.

MOLIÈRE—L'AVARE. Edited by L. M. MORIARTY, B.A., Assistant-Master at Rossall. 1s.

MOLIÈRE—LE BOURGEOIS GENTILHOMME. By the same Editor. 1s. 6d.

RACINE—BRITANNICUS. Edited by EUGÈNE PELLISSIER, Assistant-Master in Clifton College, and Lecturer in University College, Bristol. [*In preparation*

SCENES IN ROMAN HISTORY. SELECTED FROM FRENCH HISTORIANS. Edited by C. COLBECK, M.A., late Fellow of Trinity College, Cambridge; Assistant-Master at Harrow. [*In preparation.*

SAND, GEORGE—LA MARE AU DIABLE. Edited by W. E. RUSSELL, M.A., Assistant Master in Haileybury College. 1s.

SANDEAU, JULES—MADEMOISELLE DE LA SEIGLIÈRE. Edited by H. C. STEEL, Assistant Master in Wellington College. [*In the press.*

VOLTAIRE—CHARLES XII. Edited by G. E. FASNACHT. [*In the press.*

*** *Other volumes to follow.*

(See also *German Authors*, page 62.)

Masson (Gustave).—A COMPENDIOUS DICTIONARY OF THE FRENCH LANGUAGE (French-English and English-French). Adapted from the Dictionaries of Professor ALFRED ELWALL. Followed by a List of the Principal Diverging Derivations, and preceded by Chronological and Historical Tables. By GUSTAVE MASSON, Assistant Master and Librarian, Harrow School. New Edition. Crown 8vo. 6s.

Moliere.—LE MALADE IMAGINAIRE. Edited, with Introduction and Notes, by FRANCIS TARVER, M.A., Assistant Master at Eton. Fcap. 8vo. 2s. 6d.

(See also *Macmillan's Foreign School Classics.*)

GERMAN.

Macmillan's Progressive German Course.—By G. EUGÈNE FASNACHT.

PART I.—FIRST YEAR. Easy Lessons and Rules on the Regular Accidence. Extra fcap. 8vo. 1s. 6d.

Part II.—SECOND YEAR. Conversational Lessons in Systematic Accidence and Elementary Syntax. With Philological Illustrations and Etymological Vocabulary. Extra fcap. 8vo. 2s.

Part III.—THIRD YEAR. [*In preparation.*

*** *Keys to the French and German Courses are in preparation.*

Macmillan's Progressive German Readers.—By G. E. FASNACHT. First Year. [*In the press.*

Macmillan's Primary German Reading Books.
(See page 59.)

Macmillan's Foreign School Classics. Edited by G. EUGÈNE FASNACHT. 18mo.

GERMAN.

GOETHE—GÖTZ VON BERLICHINGEN. Edited by H. A. BULL, M.A., Assistant Master at Wellington College. 2s.

GOETHE—FAUST. PART I. Edited by JANE LEE, Lecturer in Modern Languages at Newnham College, Cambridge.
[*In preparation.*

HEINE—SELECTIONS FROM THE REISEBILDER AND OTHER PROSE WORKS. Edited by C. COLBECK, M.A., Assistant-Master at Harrow, late Fellow of Trinity College, Cambridge. 2s. 6d.

SCHILLER—DIE JUNGFRAU VON ORLEANS. Edited by JOSEPH GOSTWICK. 2s. 6d.

SCHILLER—MARIA STUART. Edited by C. SHELDON, M.A., D.Lit., Senior Modern Language Master in Clifton College. 2s. 6d.

MODERN LANGUAGES AND LITERATURE. 63

Foreign School Classics (German) *Continued—*
SCHILLER—WILHELM TELL. Edited by G. E. FASNACHT.
[*In preparation.*
UHLAND—SELECT BALLADS. Adapted as a First Easy Reading Book for Beginners. Edited by G. E. FASNACHT. 1s.
*** *Other Volumes to follow.*
(See also *French Authors*, page 60.)

Pylodet.—NEW GUIDE TO GERMAN CONVERSATION; containing an Alphabetical List of nearly 800 Familiar Words; followed by Exercises; Vocabulary of Words in frequent use; Familiar Phrases and Dialogues; a Sketch of German Literature, Idiomatic Expressions, &c. By L. PYLODET. 18mo, cloth limp. 2s. 6d.
A SYNOPSIS OF GERMAN GRAMMAR. From the above. 18mo. 6d.

Whitney.—Works by W. D. WHITNEY, Professor of Sanskrit and Instructor in Modern Languages in Yale College.
A COMPENDIOUS GERMAN GRAMMAR. Crown 8vo. 4s. 6d.
A GERMAN READER IN PROSE AND VERSE. With Notes and Vocabulary. Crown 8vo. 5s.

Whitney and Edgren.—A COMPENDIOUS GERMAN AND ENGLISH DICTIONARY, with Notation of Correspondences and Brief Etymologies. By Professor W. D. WHITNEY, assisted by A. H. EDGREN. Crown 8vo. 7s. 6d.
THE GERMAN-ENGLISH PART, separately, 5s.

MODERN GREEK.

Vincent and Dickson. — HANDBOOK TO MODERN GREEK. By EDGAR VINCENT and T. G. DICKSON, M.A. Second Edition, revised and enlarged, with Appendix on the relation of Modern and Classical Greek by Professor JEBB. Crown 8vo. 6s.

ITALIAN.

Dante. — THE PURGATORY OF DANTE. Edited, with Translation and Notes, by A. J. BUTLER, M.A., late Fellow of Trinity College, Cambridge. Crown 8vo. 12s. 6d.

DOMESTIC ECONOMY.

Barker.—FIRST LESSONS IN THE PRINCIPLES OF COOKING. By LADY BARKER. New Edition. 18mo. 1s.

Berners.—FIRST LESSONS ON HEALTH. By J. BERNERS. New Edition. 18mo. 1s.

Fawcett.—TALES IN POLITICAL ECONOMY. By MILLICENT GARRETT FAWCETT. Globe 8vo. 3s.

Frederick.—HINTS TO HOUSEWIVES ON SEVERAL POINTS, PARTICULARLY ON THE PREPARATION OF ECONOMICAL AND TASTEFUL DISHES. By Mrs. FREDERICK. Crown 8vo. 1s.

"This unpretending and useful little volume distinctly supplies a desideratum. .... The author steadily keeps in view the simple aim of 'making every-day meals at home, particularly the dinner, attractive,' without adding to the ordinary household expenses."—SATURDAY REVIEW.

Grand'homme.— CUTTING-OUT AND DRESSMAKING. From the French of Mdlle. E. GRAND'HOMME. With Diagrams. 18mo. 1s.

Tegetmeier.—HOUSEHOLD MANAGEMENT AND COOKERY. With an Appendix of Recipes used by the Teachers of the National School of Cookery. By W. B. TEGETMEIER. Compiled at the request of the School Board for London. 18mo. 1s.

Thornton.—FIRST LESSONS IN BOOK-KEEPING. By J. THORNTON. New Edition. Crown 8vo. 2s. 6d.

The object of this volume is to make the theory of Book-keeping sufficientl plain for even children to understand it.

Wright.—THE SCHOOL COOKERY-BOOK. Compiled and Edited by C. E. GUTHRIE WRIGHT, Hon Sec. to the Edinburgh School of Cookery. 18mo. 1s.

ART AND KINDRED SUBJECTS.

Anderson.—LINEAR PERSPECTIVE, AND MODEL DRAWING. A School and Art Class Manual, with Questions and Exercises for Examination, and Examples of Examination Papers. By LAURENCE ANDERSON. With Illustrations. Royal 8vo. 2s.

Collier.—A PRIMER OF ART. With Illustrations. By JOHN COLLIER. 18mo. 1s.

Delamotte.—A BEGINNER'S DRAWING BOOK. By P. H. DELAMOTTE, F.S.A. Progressively arranged. New Edition improved. Crown 8vo. 3s. 6d.

Ellis.—SKETCHING FROM NATURE. A Handbook for Students and Amateurs. By TRISTRAM J. ELLIS. With a Frontispiece and Ten Illustrations, by H. STACY MARKS, R.A., and Twenty-seven Sketches by the Author. Crown 8vo. 2s. 6d. (*Art at Home Series.*)

Hunt.—TALKS ABOUT ART. By WILLIAM HUNT. With a Letter from J. E. MILLAIS, R.A. Crown 8vo. 3s. 6d.

Taylor.—A PRIMER OF PIANOFORTE PLAYING. By FRANKLIN TAYLOR. Edited by Sir GEORGE GROVE. 18mo. 1s.

WORKS ON TEACHING.

Blakiston—THE TEACHER. Hints on School Management. A Handbook for Managers, Teachers' Assistants, and Pupil Teachers. By J. R. BLAKISTON, M.A. Crown 8vo. 2s. 6d. (Recommended by the London, Birmingham, and Leicester School Boards.)

"Into a comparatively small book he has crowded a great deal of exceedingly useful and sound advice. It is a plain, common-sense book, full of hints to the teacher on the management of his school and his children."—SCHOOL BOARD CHRONICLE.

Calderwood—ON TEACHING. By Professor HENRY CALDERWOOD. New Edition. Extra fcap. 8vo. 2s. 6d.

Fearon.—SCHOOL INSPECTION. By D. R. FEARON, M.A., Assistant Commissioner of Endowed Schools. New Edition. Crown 8vo. 2s. 6d.

Gladstone.—OBJECT TEACHING. A Lecture delivered at the Pupil-Teacher Centre, William Street Board School, Hammersmith. By J. H. GLADSTONE, Ph.D., F.R.S., Member of the London School Board. With an Appendix. Crown 8vo. 3d.

"It is a short but interesting and instructive publication, and our younger teachers will do well to read it carefully and thoroughly. There is much in these few pages which they can learn and profit by."—THE SCHOOL GUARDIAN.

DIVINITY.

*** For other Works by these Authors, see THEOLOGICAL CATALOGUE.

Abbott (Rev. E. A.)—BIBLE LESSONS. By the Rev. E. A. ABBOTT, D.D., Head Master of the City of London School. New Edition. Crown 8vo. 4s. 6d.

"Wise, suggestive, and really profound initiation into religious thought."—GUARDIAN.

f

Abbott—Rushbrooke.—THE COMMON TRADITION OF THE SYNOPTIC GOSPELS, in the Text of the Revised Version. By EDWIN A. ABBOTT, D.D., formerly Fellow of St. John's College, Cambridge, and W. G. RUSHBROOKE, M.L., formerly Fellow of St. John's College, Cambridge. Crown 8vo. 3s. 6d.

The Acts of the Apostles.—Edited with Introduction and Notes. By T. E. PAGE, M.A. Fcap. 8vo. [*In preparation.*

Arnold.—A BIBLE-READING FOR SCHOOLS.—THE GREAT PROPHECY OF ISRAEL'S RESTORATION (Isaiah, Chapters xl.—lxvi.). Arranged and Edited for Young Learners. By MATTHEW ARNOLD, D.C.L., formerly Professor of Poetry in the University of Oxford, and Fellow of Oriel. New Edition. 18mo, cloth. 1s.
ISAIAH XL.—LXVI. With the Shorter Prophecies allied to it. Arranged and Edited, with Notes, by MATTHEW ARNOLD. Crown 8vo. 5s.
ISAIAH OF JERUSALEM, IN THE AUTHORISED ENGLISH VERSION. With Introduction, Corrections, and Notes. By MATTHEW ARNOLD. Crown 8vo. 4s. 6d.

Benham.—A COMPANION TO THE LECTIONARY. Being a Commentary on the Proper Lessons for Sundays and Holy Days. By Rev. W. BENHAM, B.D., Rector of S. Edmund with S. Nicholas Acons, &c. New Edition. Crown 8vo. 4s. 6d.

Cassel.—MANUAL OF JEWISH HISTORY AND LITERATURE; preceded by a BRIEF SUMMARY OF BIBLE HISTORY. By DR. D. CASSEL. Translated by Mrs. HENRY LUCAS. Fcap. 8vo. 2s. 6d.

Cheetham.—A CHURCH HISTORY OF THE FIRST SIX CENTURIES. By the Ven. ARCHDEACON CHEETHAM. Crown 8vo. [*In the press.*

Curteis.—MANUAL OF THE THIRTY-NINE ARTICLES By G. H. CURTEIS, M.A., Principal of the Lichfield Theological College. [*In preparation*

Davies.—THE EPISTLES OF ST. PAUL TO THE EPHESIANS, THE COLOSSIANS, AND PHILEMON; with Introductions and Notes, and an Essay on the Traces of Foreign Elements in the Theology of these Epistles. By the Rev. J. LLEWELYN DAVIES, M.A., Rector of Christ Church, St. Marylebone; late Fellow of Trinity College, Cambridge. Second Edition. Demy 8vo. 7s. 6d.

Drummond.—THE STUDY OF THEOLOGY, INTRODUCTION TO. By JAMES DRUMMOND, LL.D., Professor of Theology in Manchester New College, London. Crown 8vo. 5s.

Gaskoin.—THE CHILDREN'S TREASURY OF BIBLE STORIES. By Mrs. HERMAN GASKOIN. Edited with Preface by Rev. G. F. MACLEAR, D.D. PART I.—OLD TESTAMENT HISTORY. 18mo. 1s. PART II.—NEW TESTAMENT. 18mo. 1s. PART III.—THE APOSTLES: ST. JAMES THE GREAT, ST. PAUL, AND ST. JOHN THE DIVINE. 18mo. 1s.

Golden Treasury Psalter.—Students' Edition. Being an Edition of "The Psalms Chronologically arranged, by Four Friends," with briefer Notes. 18mo. 3s. 6d.

Greek Testament.—Edited, with Introduction and Appendices, by CANON WESTCOTT and Dr. F. J. A. HORT. Two Vols. Crown 8vo. 10s. 6d. each.
Vol. I. The Text.
Vol. II. Introduction and Appendix.

Greek Testament.—Edited by Canon WESTCOTT and Dr. HORT. School Edition of Text. Globe 8vo. [*In the press.*

The Greek Testament and the English Version, a Companion to. By PHILIP SCHAFF, D.D., President of the American Committee of Revision. With Facsimile Illustrations of MSS., and Standard Editions of the New Testament. Crown 8vo. 12s.

Hardwick.—Works by Archdeacon HARDWICK :—
A HISTORY OF THE CHRISTIAN CHURCH. Middle Age. From Gregory the Great to the Excommunication of Luther. Edited by WILLIAM STUBBS, M.A., Regius Professor of Modern History in the University of Oxford. With Four Maps. Fourth Edition. Crown 8vo. 10s. 6d.
A HISTORY OF THE CHRISTIAN CHURCH DURING THE REFORMATION. Fourth Edition. Edited by Professor STUBBS. Crown 8vo. 10s. 6d.

Jennings and Lowe.—THE PSALMS, WITH INTRODUCTIONS AND CRITICAL NOTES. By A. C. JENNINGS, B.A.; assisted in parts by W. H. LOWE. In 2 vols. Crown 8vo. 10s. 6d. each.

Lightfoot.—Works by Right Rev. J. B. LIGHTFOOT, D.D., Bishop of Durham :—

ST. PAUL'S EPISTLE TO THE GALATIANS. A Revised Text, with Introduction, Notes, and Dissertations. Seventh Edition, revised. 8vo. 12s.

ST. PAUL'S EPISTLE TO THE PHILIPPIANS. A Revised Text, with Introduction, Notes, and Dissertations. Seventh Edition, revised. 8vo. 12s.

ST. CLEMENT OF ROME—THE TWO EPISTLES TO THE CORINTHIANS. A Revised Text, with Introduction and Notes. 8vo. 8s. 6d.

ST. PAUL'S EPISTLES TO THE COLOSSIANS AND TO PHILEMON. A Revised Text, with Introductions, Notes, and Dissertations. Sixth Edition, revised. 8vo. 12s.

THE IGNATIAN EPISTLES. 8vo. [*In the press.*

Maclear.—Works by the Rev. G. F. MACLEAR, D.D., Warden of St. Augustine's College, Canterbury, and late Head-Master of King's College School, London :—

A CLASS-BOOK OF OLD TESTAMENT HISTORY. New Edition, with Four Maps. 18mo. 4s. 6d.

A CLASS-BOOK OF NEW TESTAMENT HISTORY, including the Connection of the Old and New Testaments. With Four Maps. New Edition. 18mo. 5s. 6d.

A SHILLING BOOK OF OLD TESTAMENT HISTORY, for National and Elementary Schools. With Map. 18mo, cloth. New Edition.

A SHILLING BOOK OF NEW TESTAMENT HISTORY, for National and Elementary Schools. With Map. 18mo, cloth. New Edition.

These works have been carefully abridged from the author's large manuals.

CLASS-BOOK OF THE CATECHISM OF THE CHURCH OF ENGLAND. New Edition. 18mo. 1s. 6d.

A FIRST CLASS-BOOK OF THE CATECHISM OF THE CHURCH OF ENGLAND. With Scripture Proofs, for Junior Classes and Schools. New Edition. 18mo. 6d.

A MANUAL OF INSTRUCTION FOR CONFIRMATION AND FIRST COMMUNION. WITH PRAYERS AND DEVOTIONS. 32mo, cloth extra, red edges. 2s.

Maurice.—THE LORD'S PRAYER, THE CREED, AND THE COMMANDMENTS. A Manual for Parents and Schoolmasters. To which is added the Order of the Scriptures, By the Rev. F. DENISON MAURICE, M.A. 18mo, cloth, limp. 1s.

Procter.—A HISTORY OF THE BOOK OF COMMON PRAYER, with a Rationale of its Offices. By Rev. F. PROCTER. M.A. Sixteenth Edition, revised and enlarged. Crown 8vo. 10s. 6d.

Procter and Maclear.—AN ELEMENTARY INTRODUCTION TO THE BOOK OF COMMON PRAYER. Re-arranged and supplemented by an Explanation of the Morning and Evening Prayer and the Litany. By the Rev. F. PROCTER and the Rev. Dr. MACLEAR. New and Enlarged Edition, containing the Communion Service and the Confirmation and Baptismal Offices. 18mo. 2s. 6d.

The Psalms, with Introductions and Critical Notes.—By A. C. JENNINGS, B.A., Jesus College, Cambridge, Tyrwhitt Scholar, Crosse Scholar, Hebrew University, Prizeman, and Fry Scholar of St. John's College; assisted in Parts by W. H. LOWE, M.A., Hebrew Lecturer and late Scholar of Christ's College, Cambridge, and Tyrwhitt Scholar. In 2 vols. Crown 8vo. 10s. 6d. each.

Ramsay.—THE CATECHISER'S MANUAL; or, the Church Catechism Illustrated and Explained, for the Use of Clergymen, Schoolmasters, and Teachers. By the Rev. ARTHUR RAMSAY, M.A. New Edition. 18mo. 1s. 6d.

Simpson.—AN EPITOME OF THE HISTORY OF THE CHRISTIAN CHURCH. By WILLIAM SIMPSON, M.A. New Edition. Fcap. 8vo. 3s. 6d.

St. John's Epistles.—The Greek Text with Notes and Essays, by BROOKE FOSS WESTCOTT, D.D., Regius Professor of Divinity and Fellow of King's College, Cambridge, Canon of Westminster, &c. 8vo. 12s. 6d.

St. Paul's Epistles.—Greek Text, with Introduction and Notes.

THE EPISTLE TO THE GALATIANS. Edited by the Right Rev. J. B. LIGHTFOOT, D.D., Bishop of Durham. Seventh Edition. 8vo. 12s.

St. Paul's Epistles. *Continued—*

THE EPISTLE TO THE PHILIPPIANS. By the same Editor. Seventh Edition. 8vo. 12s.

THE EPISTLE TO THE COLOSSIANS. By the same Editor. Sixth Edition. 8vo. 12s.

THE EPISTLE TO THE ROMANS. Edited by the Very Rev. C. J. VAUGHAN, D.D., Dean of Llandaff, and Master of the Temple. Fifth Edition. Crown 8vo. 7s. 6d.

THE EPISTLE TO THE THESSALONIANS, COMMENTARY ON THE GREEK TEXT. By JOHN EADIE, D.D., LL.D. Edited by the Rev. W. YOUNG, M.A., with Preface by Professor CAIRNS. 8vo. 12s.

THE EPISTLES TO THE EPHESIANS, THE COLOSSIANS, AND PHILEMON; with Introductions and Notes, and an Essay on the Traces of Foreign Elements in the Theology of these Epistles. By the Rev. J. LLEWELYN DAVIES, M.A., Rector of Christ Church, St. Marylebone; late Fellow of Trinity College, Cambridge. Second Edition, revised. Demy 8vo. 7s. 6d.

The Epistle to the Hebrews.
In Greek and English. With Critical and Explanatory Notes. Edited by Rev. FREDERIC RENDALL, M.A., formerly Fellow of Trinity College, Cambridge, and Assistant-Master at Harrow School. Crown 8vo. 6s.

Trench.—Works by R. C. TRENCH, D.D., Archbishop of Dublin.

NOTES ON THE PARABLES OF OUR LORD. Fourteenth Edition, revised. 8vo. 12s.

NOTES ON THE MIRACLES OF OUR LORD. Twelfth Edition, revised. 8vo. 12s.

COMMENTARY ON THE EPISTLES TO THE SEVEN CHURCHES IN ASIA. Third Edition, revised. 8vo. 8s. 6d.

LECTURES ON MEDIEVAL CHURCH HISTORY. Being the substance of Lectures delivered at Queen's College, London. Second Edition, revised. 8vo. 12s.

SYNONYMS OF THE NEW TESTAMENT. Ninth Edition, revised. 8vo. 12s.

Westcott.—Works by BROOKE FOSS WESTCOTT, D.D., Canon of Westminster, Regius Professor of Divinity, and Fellow of King's College, Cambridge.

A GENERAL SURVEY OF THE HISTORY OF THE CANON OF THE NEW TESTAMENT DURING THE FIRST FOUR CENTURIES. Fifth Edition. With Preface on "Supernatural Religion." Crown 8vo. 10s. 6d.

INTRODUCTION TO THE STUDY OF THE FOUR GOSPELS. Fifth Edition. Crown 8vo. 10s. 6d.

THE BIBLE IN THE CHURCH. A Popular Account of the Collection and Reception of the Holy Scriptures in the Christian Churches. New Edition. 18mo, cloth. 4s. 6d.

THE EPISTLES OF ST. JOHN. The Greek Text, with Notes and Essays. 8vo. 12s. 6d.

THE EPISTLE TO THE HEBREWS. The Greek Text Revised, with Notes and Essays. 8vo. [*In preparation.*

Westcott and Hort.—THE NEW TESTAMENT IN THE ORIGINAL GREEK. The Text Revised by B. F. WESTCOTT, D.D., Regius Professor of Divinity, Canon of Westminster, and F. J. A. HORT, D.D., Hulsean Professor of Divinity; Fellow of Emmanuel College, Cambridge: late Fellows of Trinity College, Cambridge. 2 vols. Crown 8vo. 10s. 6d. each.

Vol. I. Text.
Vol. II. Introduction and Appendix.

Wilson.—THE BIBLE STUDENT'S GUIDE to the more Correct Understanding of the English Translation of the Old Testament, by reference to the original Hebrew. By WILLIAM WILSON, D.D., Canon of Winchester, late Fellow of Queen's College, Oxford. Second Edition, carefully revised. 4to. cloth. 25s.

Wright.—THE BIBLE WORD-BOOK: A Glossary of Archaic Words and Phrases in the Authorised Version of the Bible and the Book of Common Prayer. By W. ALDIS WRIGHT, M.A., Fellow and Bursar of Trinity College, Cambridge. Second Edition, Revised and Enlarged. Crown 8vo. 7s. 6d.

Yonge (Charlotte M.).—SCRIPTURE READINGS FOR SCHOOLS AND FAMILIES. By CHARLOTTE M. YONGE. Author of "The Heir of Redclyffe." In Five Vols.

FIRST SERIES. GENESIS TO DEUTERONOMY. Extra fcap. 8vo. 1s. 6d. With Comments, 3s. 6d.

Yonge (Charlotte M.).—(*Continued*)—

SECOND SERIES. From JOSHUA to SOLOMON. Extra fcap. 8vo. 1s. 6d. With Comments, 3s. 6d.

THIRD SERIES. The KINGS and the PROPHETS. Extra fcap. 8vo. 1s. 6d. With Comments, 3s. 6d.

FOURTH SERIES. The GOSPEL TIMES. 1s. 6d. With Comments. extra fcap. 8vo, 3s. 6d.

FIFTH SERIES. APOSTOLIC TIMES. Extra fcap. 8vo. 1s. 6d. With Comments, 3s. 6d.

Zechariah—Lowe.—THE HEBREW STUDENT'S COMMENTARY ON ZECHARIAH, HEBREW AND LXX.
With Excursus on Syllable-dividing, Metheg, Initial Dagesh, and Siman Rapheh. By W. H. LOWE, M.A., Hebrew Lecturer at Christ's College, Cambridge. Demy 8vo. 10s. 6d.

www.ingramcontent.com/pod-product-compliance
Lightning Source LLC
Chambersburg PA
CBHW020241170426
43202CB00008B/180